AF705577

Archives of Toxicology, Supplement 15

Medical Toxicology

Proceedings of the
1991 EUROTOX Congress
Meeting Held in Masstricht, September 1 – 4, 1991

Edited by the Publication Committee
H. M. Bolt, F. A. de Wolff and P. Th. Henderson

With 54 Figures and 29 Tables

Springer-Verlag
Berlin Heidelberg New York
London Paris Tokyo
Hongkong Barcelona
Budapest

H. M. Bolt
Institit für Arbeitsphysiologie an der Universität Dortmund
Ardeystraße 67, D-W 4600 Dortmund 1

F. A. de Wolff
Coronel Laboratory for Occupational and Environmental Health
Academic Medical Centre
University of Amsterdam
Meibergdreef 15
NL-1105 AZ Amsterdam

P. Th. Henderson
University of Limburg, Dept. of Toxicology,
P.O.Box 616,
NL 6200 MD Maastricht

ISBN-13: 978-3-642-77262-7 e-ISBN-13: 978-3-642-77260-3
DOI: 10.1007/978-3-642-77260-3

This work is subject to copyright. All rights are reserved, whether the whole or part of the material is concerned, specifically the rights of translation, reprinting, reuse of illustration, recitation, broadcasting, reproduction on microfilms or in other ways, and storage in data banks. Duplication of this publication or parts thereof is only permitted under the provisions of the German Copyright Law of September 9, 1965, in its current version, and a copyright fee must always be paid. Violations fall under the prosecution act of the German Copyright Law.

© Springer-Verlag Berlin Heidelberg 1992

Softcover reprint of the hardcover 1st edition 1992

The use of general descriptive names, trade names, trade marks, etc. in this publication, even if the former are not especially identified, is not to be taken as a sign that such names, as understood by the Trade Marks and Merchandise Marks Act, may accordingly be used freely by anyone.

Product Liability: The publisher can give no guarantee for information about drug dosage and application thereof contained in this book. In every individual case the respective user must check its accuracy by consulting other pharmaceutical literature.

27/3140/543210 – Printed on acid-free paper

Contents

Chronic Toxicity

Organ Toxicity

Modelling

Bioanalysis in Toxicology

Young Scientists Poster Award

Medical Toxicology

(Chair: E. Dybing, Oslo)

MEDICAL TOXICOLOGY IN EUROPE: PRESENT AND FUTURE

Kello, D.

World Health Organization Regional Office for Europe
2100 Copenhagen, Denmark

INTRODUCTION

There is no doubt that we are living in a time which could be called the "chemical age". It is estimated that about 60 000 chemicals are marketed in Europe and between 200 and 1000 new substances are introduced every year. The growth of the chemical industry shows an impressive record. It has been estimated that it has doubled in output volume every 7 - 8 years, and that the amount of chemicals produced per capita in Europe has more than doubled each decade. The primary user of synthetic chemicals in Europe, the agricultural sector, has shown an increase in application of synthetic fertilizers from 123 kg/ha in 1965 to 229 kg/ha in 1985. Moreover, it is estimated that the application of pesticides in 14 European countries has today reached about 2.71 tonnes/ha (World Resources Institute 1990). Today the chemical industry has the capacity to alter the "menu" and quantities of the various organic chemicals produced in a very dramatic fashion and at very short notice. The modern chemical age is to a large extent the result of cost-effectiveness, although many synthetically produced chemicals could be derived from natural sources.

Awareness of the wide-ranging impact of environmental pollution with chemicals and of public anxiety concerning their possible health effects, has caused all European countries to place efficient prevention and control of chemical hazards, high on the political agenda. The policy-decision process to control the risk associated with exposures to hazardous agents, also known as risk management, is an extremely complex procedure still under development as a scientific discipline. There is no generally agreed definition of risk management, but the term is used to indicate the decision-making process which combines risk assessment results with socioeconomic, technical, political and other considerations to reach a conclusion on how to control and manage exposure to suspected hazardous agents.

One of the most important activities of the WHO Regional Office for Europe is to help governments and other authorities to reduce or eliminate risk from exposure to hazardous chemicals (Guidelines for Drinking-Water Quality 1984, Air Quality Guidelines for the European Region, 1987). In line with this policy, WHO has established a broad programme focusing on the development of health-related guidelines to define exposure limits for chemicals. Thus, WHO guidelines, as recommendations for exposure limits for chemicals in food, air and drinking-water, should not be considered as standards. Standards are considered to be political decisions which should be made by national authorities in the frame of a risk management process, including the decision on acceptability of a certain risk. Because of the economic, cultural, social and political differences between European countries, it is normal and natural, for instance, that uniform drinking water or air quality standards are bot acceptable for all 32 countries. Differences in the economic resources available will force some countries to take less stringent regulatory measures, after careful prioritization of problems based on cost/benefit analysis. Therefore, WHO guidelines are considered to be an unibased scientific evaluation of all available toxicological data resulting in a quantitative risk assessment. Many WHO projects have confirmed that appropriate risk assessment relies entirely on the availability and quality of toxicological data.

The purpose of this paper is to summarize the lessons learned during the process of development of WHO recommendations for exposure limits for chemicals in food, air and drinking-water, with emphasis on particular problem areas where major improvements in risk assessment could be made through toxicology.

BASIC CONCEPT OF WHO GUIDELINES AND ROLE OF TOXICOLOGY

Risk Assessment

It is recognized that risk assessment, as practised today, must often rely upon insufficient or inadequate toxicological data. Therefore, assumptions are frequently needed in each phase of the risk assessment process to fill the various gaps in available scientific knowledge. In dealing with the uncertainties associated with the mixture of scientific facts and assumptions, it is of utmost importance to have a scientifically sound risk assessment policy in order to avoid inconsistencies in the final decisions, or to resolve points of current controversy in the scientific community. Naturally, these scientific policy statements should be changed from time to time when new scientific evidence appears.

Risk assessment applied in WHO food, drinking-water and air quality guidelines, was an organized approach giving essentially two basic answers: How likely is an event to occur? - and if it does, how bad could it be in quantitative terms? A similar approach has long been used to estimate risk associated with radioactive exposure and nuclear safety, and more recently it is being used by some agencies to estimate risk associated with toxic chemical exposures. Risk assessment applied in the establishment of WHO Guidelines consists of the following key components: exposure assessment, hazard identification (qualitative assessment), and dose-exposure assessment (quantitative assessment).

Exposure Assessment

First, the general evaluation of actual or anticipated exposure concerning the type, magnitude, time and duration was assessed, involving the characterization of a population, the number of people exposed, a profile of particularly sensitive individuals, and other specific exposure data important for the quantitative estimate of risk for any particular population.

Hazard Identification

Hazard identification was based on the evaluation of all available toxicological data to characterize the strength of evidence indicating that potential health effects might occur in an exposed human population. The decision to consider a substance as a carcinogen is based on a qualitative evaluation of all available information on carcinogenicity, thus ensuring that the association is unlikely to be due to chance alone. It was concluded that the qualitative evaluation applied by IARC should serve as the baseline in establishing the quality guidelines for carcinogens (IARC 1982), and that the IARC categorization scheme should be used if no new divergent evidence was available.

Dose Response Assessment

Dose response assessment was the process of estimating the relationsship between the dose of a substance and the incidence of an adverse health effect, particularly in the conditions of low-level long-term exposure in the general population. In the absence of data describing environmental low-dose-effects, two different approaches for dose-response characterization were used, one for "threshold" endpoints and another for "non-threshold" endpoints.

For those agents that are causing effects by a "threshold" dose mechanism - meaning no significant health risks are likely to occur until a certain dose is reached - the safe levels were defined by applying safety factors to the no-observed-adverse-effect level (NOAEL) or to the lowest-observed-adverse-effect levels (LOAEL) in human or animal bioassay studies; human data were used whenever possible.

For "non-threshold" effects (e.g., genotoxic carcinogenic effects), meaning that there may be some risks associated with any exposure, dose-response extrapolation models were used to estimate cancer risks associated with low dose exposure based on observed incidence in humans or animals exposed to high doses. The last step in the risk assessment process combines the exposure and effects assessment, describing in quantitative terms the nature and magnitude of risk along with the uncertainties associated with the entire process of risk assessment.

PROBLEMS RELATED TO AVAILABILITY OF APPROPRIATE TOXICOLOGICAL EVIDENCE

Ideally, guideline values should represent concentrations of chemical compounds in environmental media that would not pose any hazard to the human population (Air Quality Guidelines for the European Region 1987). However, the realistic assessment of human health hazards necessitates a distinction between absolute safety and acceptable risk. To aim at achieving absolute safety, one would need a detailed knowledge of dose-response relationships in individuals in relation to all sources of exposure, the types of toxic effect elicited by specific pollutants or their mixtures, the existence or nonexistence of "thresholds" for specified toxic effects, the significance of interactions and the variation in sensitivity and exposure levels within the human population. However, such comprehensive and conclusive data on environmental contaminants are not always available. Very often the relevant toxicological data are scarce and the quantitative relationships uncertain; scientific judgement and consensus therefore play an important role in establishing acceptable levels of population exposure. Value judgements are unavoidable, because terms such as "adverse" and "sufficient evidence" are not in themselves totally objective, their meaning being based on generally agreed judgements.

Safety Factors and Scientific Uncertainties

The single most difficult task to be accomplished in the risk assessment of chemicals is the question of uncertainties. This problem was solved in the past by regulatory agencies using so-called safety factors. Safety factors were usually applied to derive exposure limits based on accepted criteria for adverse effects on health. The rationale has been that such a factor allows for a variety of uncertainties, for example about possible undetected effects on particularly sensitive members of the population, synergistic effects of multiple exposure, and the adequacy of existing toxicology data.

Traditionally, the safety factor has been used to allow for uncertainties in extrapolation from animals to humans, from high dose laboratory experiments to the low levels of environmental exposures, and from a small group of individuals to a large population. Most of the factors which are taken into account in deciding the uncertainty factor or the margin of protection can be grouped under the heading of scientific uncertainty.

The basic philosophy underlying this problem is that uncertainty pervades the entire health risk assessment process regarding chemicals because the "premarketing testing" can only reduce the range of possibilities and cannot predict the final outcomes in the real world with absolute certainty, which includes all those risks from chemicals and their interactions in the environment which are not detected prior to human exposure. Since nowadays the regulation of chemicals control relies exclusively on premarketing testing, there is little or no opportunity to watch for unforeseen effects of a chemical which is gradually being introduced into the environment. It must be emphasized that the real impact of the modern chemical age is the capacity to change direction almost overnight. This raises the issue of the role of proper premarketing testing to one of utmost importance. However, the toxicological information that could become available once a chemical was on the market, which is usually obtained through clinical toxicology or epidemiological studies, is of equal importance in correcting inappropriate decisions based on insufficient data provided by premarketing studies.

Reducible Uncertainties and Premarketing Testing

As indicated before, every effort should be made to reduce uncertainties related to premarketing testing. First of all, international agreements such as the OECD guidelines for testing and GLP should be fully endorsed (OECD Guidelines for Testing for Chemicals 1981). There are still many scientific laboratories which are embarking on so-called toxicological research which do not follow the above-mentioned guidelines, assuming that it is not necessary because their results will be used only for scientific purposes and not for the licensing of a particular chemical. Once such results are published in the open literature there is, for obvious reasons, no remark that the GLP principles were not followed, and therefore in most cases the data from this type of studies are used for the risk assessments.

The small size of the test sample is the next uncertainty which introduces the bias, which this smallness has on the expected health outcome, because the sample group may not be typical of the population from which it was selected, and even less so when compared to the ultimately exposed population.

Furthermore, there is an uncertainty which remains because of the specific conditions of the toxicological study, and the inability to generalize from these limited conditions to the range of conditions which exist in the real world.

The number of repeated studies reduce the uncertainty each time when a test is repeated on a different subject or under different conditions, although uncertainties cannot be reduced to zero unless the tests are repeated an infinite number of times. Therefore, some reducible uncertainty must always remain and the extent depends on the number of studies which are conducted. This point is important in balancing the costs of reducing uncertainty against the benefit from further reduction. In essence, given the natural variety of individual reactions to a given dose of a given substance, it is to be anticipated that a small sample of exposed animals or people might have reactions very different from that what would actually occur throughout the entire population. This is because the small group might by chance be made up of a number of individuals who are either unusually susceptible or unusually tolerant to exposure to particular chemicals.

A very usual problem of uncertainty is caused not by the size of the sample but rather by the design of the test. Even if the study was performed under the GLP conditions, experience shows that very often the premarketing study was not able to observe very specific health effects caused by exposure to a particular chemical. This is the problem of so-called "false-negative" studies which underline the rule that failure to produce a significant health effect is not sufficient to prove safety, but proves only that within the limits of the experiment no risk was observed. A large number of experimental procedures may be successful in discovering an earlier non-observed health effect at a particular dosage and it is obvious that successive procedures will narrow this ancertainty.

The number of combinations of factors which might influence the health outcomes of exposure to chemicals is nevertheless infinite, which means that the uncertainty from this source can never be entirely eliminated either. Theoretically it would mean that at a given dosage it would be necessary to vary all of the various combined inputs through air, diet, light, activities, etc. across the whole of their possible ranges. Again theoretically, it would mean that the "no-observed health effects" cannot be accepted with certainty until all of these combinations have been carried out. However, all that implies that the aggregate costs of eliminating all uncertainties from this source would be infinite. It is fundamentally important that the <u>correct amount of testing</u> of chemical substance occurs prior to its mass production. Current experience suggests that most of the chemicals produced in high quantities proceed through about six years of testing at a cost of about US 1 million prior to the release in Europe. However, it is important to underline that there are about 64 000 previously marketed chemicals which have never been seriously tested. Also it is assessed that despite all these efforts, not more than 2000 chemicals from the OECD priority list will have been properly studied by the year 2000.

Irreducible Uncertainties

There are some forms of uncertainties which are never reduced irrespective of the amount of premarketing studies performed, which could be considered as the irreducible source of uncertainty. In principle there are some health effects which could not be observed under existing methodologies until actual human exposure occurs. There are three main related aspects of uncertainty:

First, the uncertainty coming from extrapolation from laboratory animals and other species to humans;

Second, the uncertainty introduced in extrapolating from high experimental dosages to the low levels of environmental exposures;

Third, the uncertainty caused by extrapolation from short-term experimental exposures to the long-term lifespan of humans.

For practical reasons it is not really possible to perform animal studies testing for long-term exposure to low doses of chemicals. Since most testing occurs over much shorter spans, with much greater daily doses, all those studies provide only a rough basis for comparison with human long-term exposure to chemicals. It is obvious that those uncertainties are inherited into the capacity of toxicological science and there is little to be done about reducing this form of uncertainties related to premarketing experiments under laboratory conditions.

However, the above-mentioned main sources of irreducible uncertainty should not be understood as a license for the decision-makers to do nothing. One solution to the above-mentioned problem would be the phased introduction of registered chemicals on the market. It would also assume the introduction of well organized surveillance of populations potentially exposed to a particular chemical following the modern principles of medical toxicology. This proposal follows the principle that there is no substitute for human data for reducing the uncertainties. The cost of acquiring this type of data could be minimized after a substance had successfully completed premarketing testing, similar to the studies that have been performed in the case of pharmaceuticals. The proposed approach should be properly evaluated and rationalized in order to be applied to mass produced chemicals. A proper record-keeping practice in clinical toxicology related to occupational exposure during the early stage of chemical production, could also play a very important role in picking up early indications of potential health effects. Specific legislation related to the privacy of data in most European countries represent a major obstancle in this respect.

GENETIC SUCEPTIBILITY TO CHEMICALS - NEW CHALLENGE IN MEDICAL TOXICOLOGY

Subgroups of the population are particularly susceptible to the adverse effects of certain chemicals due to their genetic constitution, although the public health implications have so far been limited (WHO 1991). Genetic predisposition to several diseases is now becoming more clearly defined, but individual genetic predisposition is only one of the many factors affecting individual responses to chemicals. This makes the whole task of identifying and protecting high-risk populations very complex. The current regulatory practice for setting health-related environmental or occupational limits for hazardous chemicals does not specifically take account of the existence of groups in the population who are hypersusceptible owing to genetic differences. Existing regulation in most countries is restricted to protecting groups such as young people and women. The conventional application of uncertainty/safety factors when setting standards cannot fully guarantee the protection of hypersusceptible individuals. The detection of variations in individual susceptibility to react to chemical exposure raises dilemmas about serious ethical issues, such as equal rights to enjoy opportunities in soceity, equal rights in free enterprises, and rights to privacy, individual freedom and good health.

The existence of genetic polymorphisms for the metabolism of foreign compounds has been amply demonstrated. The resulting differences in enzyme activity from person to person, sometimes of several hundredfold, explain the existence of subjects who are hypersusceptible to the effects of both drugs and industrial chemicals. Genetic factors may either predispose people to suffer from the adverse effects of environmental exposure or protect them against such effects. The diseases that are influenced by such genetic factors include atherosclerosis, various malignancies, allergic and other immunological disorders, as well as specific intoxications.

The quantitative measurements of chemical adducts with DNA and proteins is now possible; such methods indicate exposure to certain chemicals. These methods also show the range of variability between individuals in metabolic activation and in DNA repair. Animal models of genetic susceptibility to environmental chemicals are useful tools to illuminate the mechanisms involved, but are of limited use in predicting genetic predisposition in humans. Although taking account of age, gender, nutrition, lifestyle and existing disease, both epidemiological studies and regulatory practice have not sufficiently addressed the significance of genetically hypersusceptible subgroups.

When adequate and sufficient information exists, detection of the genetic characteristics that influence people's susceptibility to environmentally induced diseases may serve to guide intervention, and improve the health and quality of life of the affected individual or population.

It is therefore recommended that research in this area should take new directions, including:

- large prospective cohort studies to address genetic host susceptibility, particularly for polymorphic genes that govern the variable activation and detoxification of chemicals;

- familiy studies to characterize the mode of inheritance of genetic predisposition;

- improvement of methods to determine macro-molecular adducts and other biological markers in the same individual; these methods should be used to detect and characterize hypersusceptible subgroups;
- animal models need to be developed and refined to obtain a better understanding of genetic susceptibility to chemical exposure; and
- basic research in statistical modelling and risk assessment methods to assess the risks of affecting the health of hypersusceptible individuals.

Research programmes need better coordination, particularly between different countries, to improve the efficient use of resources and the harmonization of procedures, and to establish joint cell and tissue banks from well defined populations. WHO will explore its role in this area. New sources of funding should be made available by governments and other sources to ensure the satisfactory conclusion of such studies.

REFERENCES

World Resources Institute (190) World Resources 1990-91. Basic Books, New York

WHO (1987) Guidelines for Drinking-Water Quality. Vol. 1, Recommendations. WHO, Geneva

WHO (1987) Air Quality Guidelines for the European Region. WHO Regional Publications, European Series No. 23, WHO Regional Office for Europe, Copenhagen

Anderson EL (1985) Quantitative approaches in use in the United States to asses cancer risk. In: Vouk VB et al. (eds) Methods for estimating risk of chemical injury: human and nonhuman biota and ecosystems, Scope 26, John Wiley & Sons, New York, pp 405-436

IARC (1982) Chemicals, Industrial Processes and Industries Associated with Cancer in Humans. IARC Monographs on the Evaluation of the Carcinogenic Risk of Chemicals to Humans, Vol. 29, Suppl. 4, Lyon

OECD (1981) OECD Guidelines for Testing for Chemicals. Organisation for Economic Co-operation and Development, Paris

WHO (1991) Ecogenetics - Genetic Predisposition to the Toxic Effects of Chemicals, WHO Regional Office for Europe, Chapman & Hall, London

MEDICAL TOXICOLOGY - CLINICAL ASPECTS

Vale, J. A.

West Midlands Poisons Unit, Dudley Road Hospital, Birmingham, B18 7QH, U.K. and Section of Therapeutics and Clinical Toxicology, Department of Medicine, University of Birmingham, B15 2TT, U.K.

In a recent review, Dayan (1989) has written, "As so much toxicological work involves decisions about human health, as well as sophisticated laboratory and animal experimentation of types common to all biomedical sciences, it might be expected that many toxicologists would be physicians." However, Dayan's survey of the membership of the British Toxicology Society indicates that only 1-2% of members are medically qualified.

While I believe that the possession of a medical qualification and of impressive clinical skills is generally of importance and help to the medical toxicologist, such a qualification is not mandatory and there are a number of medical toxicologists who have made impressive contributions to toxicology who are not medically qualified. Notwithstanding this, the possession of appropriate medical knowledge and clinical experience is of paramount importance if, for example, a patient's clinical features, which may or may not be related to toxic exposure, are to be interpreted correctly.

The organisers of this meeting have helped participants by indicating their own understanding of the term "medical toxicology". While agreeing with their general strategy, I would like to expand the model further indicating, firstly, those areas of knowledge and influences which most guide the medical toxicologist in his day-to-day work (Fig.1) and, secondly, those areas of toxicology and related practice which are the concern, in part at least, of the medical toxicologist.(Fig.2)

In addition to clinical skills (Fig.1), the medical toxicologist must also be able to apply relevant kinetic and biochemical data so that treatment strategies can be developed and refined, based on the likely mechanisms of toxicity. In addition, an understanding of pathological appearances is necessary if the results of toxicity tests are to be extrapolated accurately to man and the results of forensic examinations are to be interpreted correctly. The medical toxicologist will also need to recognise the limitations of analytical methods, both as regards their sensitivity and specificity.

The increased public awareness and concern regarding the potential effects of pesticides and other chemicals means that the medical toxicologist must be familiar with the occupational and environmental

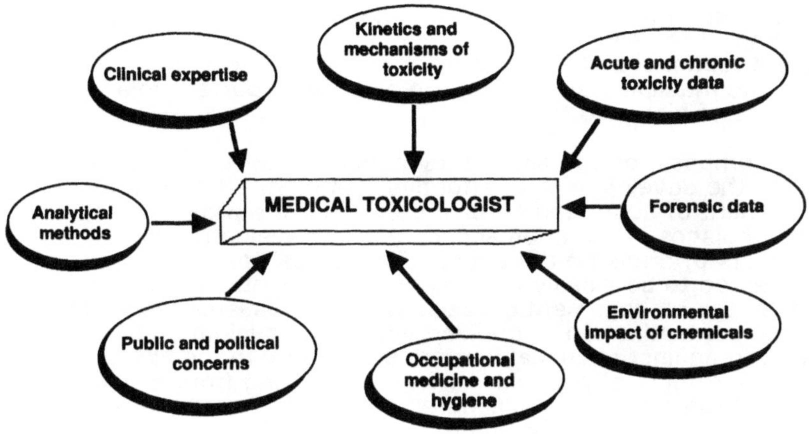

Fig 1. Areas of knowledge and influences which guide the medical toxicologist.

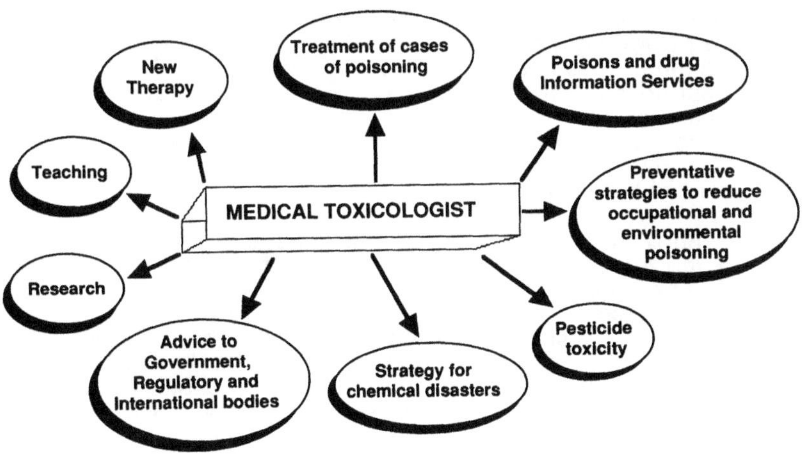

Fig 2. Areas of toxicology which are the concern of the medical toxicologist

impact of a wide range of agents. Thus, a detailed knowledge and understanding of the clinical, biochemical, kinetic, analytical, forensic, occupational and environmental aspects of toxicology is essential if the medical toxicologist is to play the important role now expected of him in today's society.

This will encompass, of course, the traditional therapeutic role (Fig. 2), including the development of effective treatment strategies for the management of acute and chronic poisoning, the provision of expert advice via a poisons information service and the evaluation of the adverse effects of drugs on the body. The medical toxicologist can also be expected to be closely involved in the development of strategies for the management of major chemical disasters including the chemical contamination of drinking water, the evaluation of antidotes used against chemical warfare agents and in the assessment of the adverse effects of pesticides whether resulting from a single exposure or chronic low-level exposure. Advice to Government and to Regulatory and International bodies is an increasingly important role for the medical toxicologist, but of even greater scientific significance is the organisation of training and research in toxicology.

CLINICAL TOXICOLOGY

Poisoning due to accidental or deliberate ingestion or inhalation of drugs or chemicals is a common, acute medical emergency. Patients who are poisoned have special physical and socio-psychiatric problems that demand attention, though the majority of cases can be managed safely and effectively in acute medical wards at local hospitals. However, specific and specialised methods of treatment are required in some cases of severe poisoning if permanent disability or death are to be avoided. Yet, even severe cases of poisoning can be usually be managed by local hospital physicians if they are given the appropriate "high level" clinical advice. Occasionally, it is necessary to transfer severe cases of poisoning to a regional or national centre, if local expertise and facilities are inadequate to meet the clinical challenge presented by the patient.

Consequently, many medical toxicologists are based in a Poisons Unit/Centre. Such toxicologists will have a broad training in internal medicine and a specialist training in toxicology. In addition to providing expert advice to colleagues via a Poisons Information Service, such staff will often be clinically responsible for the management of patients suffering from the effects of acute or chronic poisoning and the detoxification of substance abusers. It is my contention that a first-rate Poisons Information Service can only be provided if there is a clear link between the provision of information and the delivery of clinical care. In this respect I concur with Goulding (1991) that "Only in this way will the members of such units be respected for their professional competence and authority and be capable of furnishing reliable opinions... as well as being adept at differential diagnosis which can be so crucial."

Poisons Units need the support of specialised laboratory facilities not only to assist with the diagnosis of cases of poisoning but also to increase understanding of mechanisms of toxicology and to develop new therapies and to reassess the role of traditional ones. Medical toxicologists will also be involved in the vetting of analytical requests and the interpretation of the results.

An integrated service encompassing expert advice, dedicated treatment facilities and a specialised laboratory has important benefits for teaching, training and research in clinical toxicology. This pattern has been followed closely in Birmingham. The West Midlands Poisons Unit and the associated Regional Laboratory for Toxicology provide:

1. **Dedicated beds** for the management of poisoned patients. The Unit acts as a focus for the provision of expert medical, psychiatric and social care. The Unit is staffed by physicians with a special interest in, and experience of, medical toxicology. A Consultant Psychiatrist with a specific commitment to the Poisons Unit provides psycho-social assessment on a daily basis.

2. **A Poisons Information Service.** Information is given to medical practitioners and nursing staff, and occasionally to veterinary practitioners (but not to members of the public) on the ingredients and toxicity of the many drugs, chemicals, household products and plants available in present-day society. This service is available 24-hours-a-day, 365-days-a-year and at all times physicians with extensive training in medical toxicology are available to provide expert advice to enquirers.

3. **Expert clinical advice** to physicians on the management of specific cases of poisoning. In such cases all the available information, circumstantial, clinical, laboratory and from other sources has to be assessed in relationship to the substances and circumstances of exposure to determine a detailed plan for clinical management which will maximise the chances of the victim surviving and minimising the risk of short- and long-term sequelae. This expert advice must be supplemented by a comprehensive collection of reference works and original papers.

   If necessary and appropriate, severely poisoned patients are transferred from other hospitals for more intensive and specialised treatment which cannot be provided locally even after advice from the Poisons Unit.

4. **An out-patient advisory service.** The Unit has for some years offered an advisory service particularly on chronic medical problems which are alleged to have a toxicological basis; many of these have an occupational origin.

5. **Analytical support** both for the Poisons Unit and the surrounding Region. A close working relationship exists between the medical and analytical toxicologists which strengthens and improves the quality of the overall service even further.

TOXICITY OF PESTICIDES

Acute pesticide poisoning is an important cause of worldwide morbidity and mortality. It has been estimated that there are three million cases of acute **severe** pesticide poisoning each year with 220,000 deaths (WHO 1986, 1990). Ninety-five per cent of fatal pesticide poisonings occur in developing countries, where 25 million episodes of intoxication occur annually among agricultural workers alone (Jeyaratnam 1985, 1990). However, the problem is not restricted to developing countries as between 150,000 and 300,000 pesticide-related illnesses are thought to occur each year in the USA (Coye 1985); data for Europe are likely to be of a similar order.

As a result, there is increasing media coverage reflecting the growing public concern over the potential toxicity of pesticides. These fears need to be addressed adequately for two reasons. Firstly, no segment of the general population is completely protected against exposure to pesticides and secondly, the public are now much more knowledgeable about the toxicity of pesticides than in the past due to a variety of publications aimed at this market (Lang and Clutterbuck, 1991; Robbins, 1991). These and other texts provide basic information for the non-specialist so that the interested reader may be even better informed about pesticides than the doctor with whom he discusses his symptoms! Furthermore, the formation of patient groups in many countries, such as the Pesticide Exposure Group of Sufferers (PEGS) in the UK, indicates and supports the need for careful clinical assessments to be undertaken by a medical toxicologist of those individuals claiming to be suffering from the on-going toxic effects of exposure to a pesticide. Such an assessment is important both for the individual and in epidemiological terms. Many of these "sufferers" complain of difficulty in reporting their case to official agencies and of finding an informed doctor who is interested in this type of clinical problem. There is, therefore, a need for suspected cases of this kind to be audited clinically so that symptoms due to pesticide exposure can be distinguished from those which have a different aetiology. Such clinical assessments must, of necessity, take note of all available evidence, such as the recent confirmation (Rosenstack et al 1991) that even a single episode of clinically significant organophosphorus insecticide intoxication may be associated with a persistent decline in neuropsychological functioning; this finding has considerable implications both for the developing and developed nations.

The WHO (1990) has recommended that "every effort should be made to perform epidemiological studies "and to improve understanding of the effects of pesticides on the public health. In order to obtain data on pesticide exposures and suspected adverse health effects, the UK Health and Safety Executive, with the encouragement of the Department of Health and the British Medical Association, has funded the West Midlands Poisons Unit to manage a pilot surveillance scheme involving the use of a "green card" (Booth et al 1991). This scheme is analogous to the reporting system used in many countries for the recording of adverse drug reactions. Each doctor returning a "green card" is contacted promptly and guidance is given about the possible connection between the patient's symptoms and exposure to a pesticide. In addition, all relevant national data on the toxicity of pesticides is being collated and critically assessed. The scheme will also involve a detailed clinical audit of referred cases which should distinguish patients with pesticide poisoning from reported incidents which are not confirmed by clinical follow-up.

CHEMICAL DISASTERS

Most chemical accidents are minor but occasionally there is a disaster such as Bhopal in India in 1984 when the release of methyl isocyanate left some 2,500 dead and many more, perhaps 50,000, temporarily or permanently disabled. Most developed countries have the resources, infrastructure and the established mechanisms to respond adequately to chemical emergencies, so that the population can be evacuated rapidly from the affected area, victims can be resuscitated and transferred speedily to hospital and fires can be brought under control without causing further harm.

However, sensational reporting by the media may produce alarm and panic even when the emergency services perform their allotted tasks competently. Thus, a medical toxicologist may need to act in three ways. Firstly, the medical toxicologist should provide reassurance in the early stages or, if that is inappropriate, an accurate assessment of the likely risk to the population. Secondly, he may need to advise on the best therapy or even co-ordinate treatment for a large number of casualties. Thirdly, after the disaster is under control, he may be called upon to assess the likelihood of late sequelae.

In our own Health Region there is an Advisory Group on Chemical Hazards with a supporting Task Force, which is activated, when appropriate, by the Regional Director of Public Health (Harrison et al 1991). The core group of experts includes two medical toxicologists, two occupational physicians and two public health physicians. A recent incident investigated by the Task Force involved a fire at a factory; a tank containing dichloromethane (methylene chloride), methanol and sodium hydroxide caught fire. The local radio announced that mustard gas and phosgene had been released and although this was discounted by a medical toxicologist at an early stage, a large number of "casualties" were taken to the local hospital; these included a large number of emergency personnel who were told to have a check-up. Subsequently, supposed "experts" claimed that the exposed population would suffer an increased carcinogenic risk and that the incidence of fetal abnormalities would be increased in those pregnant at the time. In reality, a comprehensive investigation showed that immediate symptoms in the casualties, if present, were mild and were probably unrelated, in the majority of cases, to the fire. No significant delayed symptoms occurred and, in the circumstances prevailing no long-term sequelae were to be expected.

TRAINING IN MEDICAL TOXICOLOGY

Dayan (1989) has investigated the teaching of toxicology in the 24 UK medical schools. Two of the 24 schools did not teach any toxicology, which he defined as, "the study of the harmful effects on man of exposure to chemicals and medicines at work, at home, in food and in the environment." The number of hours given to teaching toxicology in the complete medical course at other medical schools ranged from 1 to 12 hours (median 5 hours; mode 8 hours). Dayan concluded, "There is, however, a fairly consistent pattern of minimal attention to toxicology in general and little if any formal coverage of it outside the area of pharmaceuticals... I must ask whether the medical profession, in overlooking toxicology, is avoiding an area of scientific excitement, of social and economic action and of public anxiety."

I would strongly support Dayan's contention that the teaching of toxicology to medical undergraduates should be given a greater priority. Realistically, however, this will be difficult to achieve in the short-term for two reasons. Firstly, the medical curriculum is already overloaded in most Schools and Faculties and secondly, there is a dearth of committed medical toxicologists to undertake the task even if time was made available. Without overlooking the importance of medical undergraduate teaching, it may at present be more productive for medical toxicologists to concentrate on medical (and non-medical) postgraduates.

Our own Unit contributes to medical undergraduate training in toxicology by organising a series of lectures and case discussions; the case discussions, which involve the active participation of six students who "discuss the case" before an audience composed of the whole undergraduate year, are particularly popular. In addition, regular major symposia lasting 4-5 days on recent advances in clinical toxicology attract an international audience. We also organise one of the five modules on an MSc in Toxicology course which is taken by approximately twenty, usually non-medical graduates, each year. Shorter, 1-2 day courses for occupational and emergency physicians are also held regularly.

CONCLUSION

Medical toxicologists believe that toxicology is an area of scientific excitement and of social and economic relevance. In the main, medical toxicologists have risen to the challenges presented to them over the last 30 years and now the worth of these endeavours is being recognized by others - medical toxicology has been placed on the political agenda, both nationally and internationally. For example, the recent resolution by the European Council (1990) to improve the prevention and treatment of acute human poisoning has immediate relevance to our work. Despite our relatively small numbers, medical toxicologists are already making a significant contribution to toxicology but we now have new opportunities to address, particularly within the wider fields of public and environmental health. By working together, I have every confidence that we will meet these opportunities and challenges.

Acknowledgement

The advice of Dr T.J. Meredith, Consulting Toxicologist, West Midlands Poisons Unit, is gratefully acknowledged.

REFERENCES

Booth C, Morgan D, Lamont T, Vale J A (1991)
Doctoring the Environment: Pesticides 'Green Card' Surveillance Scheme. Br Med J (to be published)

Coye M J (1985)
The health effects of agricultural production. I. The health of agricultural workers. J Public Health Policy 6: 349-70

Dayan A D (1989)
Who needs toxicology? J R Soc Med 82: 320-22

EC (1990)
Official Journal of the European Communities C329:6-18

Goulding R (1991)
Poisons Information Services: A look ahead. Hum Exp Toxicol 10: (to be published)

Harrison M, Aw T C, Krishnan G, Jones A F, Vale J A (1991)
Major chemical disasters. Br Med J 302: 657

Jeyaratnam J (1985)
1984 and occupational health in developing countries. Scand J Work Environ Health 11: 229-34

Jeyaratnam J (1990)
Pesticide Poisoning: A major global health problem. World Health Stat Q 43: 139-44

Lang T, Clutterbuck C (1991)
P is for Pesticides. Ebury Press. London

Robbins C (1991)
Poisoned Harvest. Victor Gollancz. London

Rosenstack L, Keifer M, Daniell W E, McConnell R, Claypoole K (1991)
Chronic central nervous system effects of acute organophosphate pesticide intoxication. Lancet 338: 223-27

World Health Organization (1986)
Informal consultation on planning strategy for the prevention of pesticide poisoning. Geneva 25-29 November 1985. WHO/VBC/86.926. WHO, Geneva

World Health Organization (1990)
Public health impact of pesticides used in agriculture. WHO, Geneva

RESEARCH ASPECTS IN CLINICAL TOXICOLOGY

Dag Jacobsen, M.D., Ph.D.
Section of Acute Medicine & Cardiology
Department of Medicine
Ullevål University Hospital

Clinical toxicology is an old discipline with a young an still immature academic impact. Over the last two decades, however, much research information has been available with better and better scientific quality. Though most of the documentation is still based on case reports, their quality has improved and some multicenter studies have also been performed in which different treatment procedures have been evaluated.

The basis for documentation in clinical toxicology can be classified as follows:

- Clinical studies

- Case reports

- Animal experiments

- Pharmacological studies

Clinical studies

The clinical studies are mainly dealing with epidemiology and evaluation of efficacy of treatment procedures. The majority of the epidemiological studies are unfortunately not of good quality. Many studies are retrospective and based on patient records with obvious limitations as to their interpretation. If prospective in design, many studies deal only with selected groups of patients, e.g. from intensive care units, without attention to the population size in that area. Comparision of epidemiological parameters such as incidence of acute poisonings is therefore impossible. Some studies have, however, related the patient cohort to the local population allowing for direct comparisions between studies (Rygnestad, 1988; Jacobsen et al., 1984).

Such patient cohort can also be followed over years to provide valuable information as to the follow-up of self-poisoned patients. A critical evaluation of this follow-up is important as the 5-year mortality for this patient cohort has been shown to be as high as 13% (Ekeberg et al., 1991). Those admitted for suicidal attempts or parasuicides tend to commit suicide later. Therefore special attention should be paid to the follow-up of these patients.

Few double-blind placebo-controlled clinical studies have been performed within clinical toxicology. This may require a multicenter design and the cooperation of many centres in different countries in the actual type of poisoning is rare. In poisonings that are frequently occuring, such as with the benzodiazepines or opiates, studies of efficacy of specific treatment procedures may, however, be performed within one department. Such studies have been undertaken for the new effective antidotes such as naloxone, flumazenil and anti-digoxin. The new technique of producing monoclonal antibodies for use as antidotes and antivenoms will lead to several clinical studies of which some are already ongoing.

The dramatic efficacy of naloxone and flumazenil and the involvement of endogenous opiate and benzodiazepine systems, may explain the uncritical overuse of these antidotes in situations where their efficacy has never been proven, e.g. in ethanol intoxicated patients.Therefore, properly designed studies have also been necessary in order to demonstrate where these antidotes are not efficacious.

Bismuth (1986) treated patients for benzodiazepine poisoning in a double blind randomized study, injecting one single dose of placebo or flumazenil. Two of the 20 placebo subjects partially awoke versus 17 of the 20 patients given flumazenil, thus clearly documenting the effect of the antidote. Later similar studies have confirmed this beneficial effect (Aarseth et al, 1988; Ritz et al., 1990).

There have been evidence that central nervous effects of ethanol are mediated through the GABA-system. This evidence have also been supported by 'uncontrolled' case reports. Such information may lead to uncritical overuse of flumazenil in acute poisonings. Therefore the effect of flumazenil on psychometric performance was studied in 8 healthy volunteers intoxicated by ethanol (Clausen et al., 1990). Flumazenil did not improve psychomotor functions in these intoxicated subjects and therefore has no role in the treatment of ethanol intoxication.

It has also been suggested that ethanol may exert some of its effects via the endogenous opiate system, as illustrated by the studies by Jefferys et al. (1980) and Jeffcoate et al. (1979) where naloxone was found to antagonize some of the ethanol effects. These studies were also supported by anecdotal case reports. These results lead also to an overuse of naloxone in ethanol intoxicated patients. However, Nuotto et al.(1984) demonstrated in their placebo-controlled double-blind study that naloxone had no effects on the signs and symptoms of ethanol intoxicated patients.

Retrospective studies should have no place in the documentation of various treatment regimes. In these studies, an uncritical collection of case reports is usually compared with historical 'controls'. An example is the often recommended use of dimercaprol (BAL) in lead poisoning in children. The basis for this documentation is patient records of children treated with dimercaprol. The control group consisted of children treated 10-15 years earlier when dimercaprol was not available. In general, such patient group are not comparable, and beneficial effect of using dimercaprol in lead poisoning has neither been documented in animal models. In the latter experiments even a harmful effect of dimercaprol has been demonstrated.

Case reports

Clinical toxicology cannot exist without the information obtained from well performed case studies. These results must, however, be interpreted with caution as there is no control group and most patients do well on symptomatic treatment only. If these facts are not considered, the case report literature may be correctly called 'the nightmare of science'. In most situations, the best conclusion from case resports would be that the use of e.g. an antidote was associated with a favourable outcome. In some cases, however, the effect of antidote administration is so convincing that the efficacy may not be questioned. The use of cobalt-EDTA in cyanid poisoning and naloxone in opioid poisoning are such examples.

The case reports are valuable becaue they often represent extreme situations in our own species. Furthermore, many poisonings are rare and clinical studies are therefore not possible from a practical point of view. In some poisonings, e.g. with methanol, there are significant differences in toxicity between species (Jacobsen & McMartin, 1986; Tephly & McMartin, 1984). Relevant information therefore hase to be derived from human case reports or studies in primates.

Animal experiments

The major advantage of the animal model is the controlled experimental conditions. Such studies also provide valuable information about toxicity through the widely used and accepted LD_{50}-values. In clinical toxicology, the animal model has been very useful in elucidating the mechanism of paracetamol toxicity and the evaluation of antidotes (chelating agents) in heavy metal toxicity. A major problem in the latter type of studies is to define the best efficacy parameter in evaluating the effect of the antidote. Should it be

- increased excretion of heavy metal?
- reduced mortality?
- reduced concentration of toxic agent in target organs?

- reduced concentration of toxic agent in critical organ?
- other sophisticated methods?

Most toxicologists would agree in that increased excretion of heavy metal caused by antidote per se is not sufficient to document effect. The pragmatic clinical toxicologist would require a significant reduction in mortality in the antidote treated group in order to recommend clinical use of the chelating agent in question. Though a controversial issue, many clinical toxicologists would also accept reduced concentration of toxic agent in critial organ as an efficacy parameter provided good therapeutic index of antidote.

An axample of very useful animal data is the classical Chinese studies performed several decades ago demonstrating the efficacy of dimercapto-succinic acid (DMSA) in various heavy metal poisonings in mice (Ding et al.; 1965). In these studies a reduced mortality was used as efficacy parameter. These extensive studies have lately been reviewed in English by Ding & Liang (1991).

Pharmacological studies

In clinical toxicology, limited information is obtained from pharmacological studies. In this situation, one extrapolates the side effects of the drug under therapeutic conditions into the overdose situation. It may be true that drugs causing drowsiness in 20% of users will cause coma in the overdose situation, but such extrapolation should be done with caution.

The traditional 'phase I and II-studies' also have to be performed for most antidotes used in clinical toxicology. If there exists a commercial market for the new antidote, e.g. naloxone and flumazenil, these studies will be undertaken by the drug companies themselves. However, in the 'orphan drug' situation where the earnings will not cover the development expenses, the clinical toxicologists may have to do these studies themselves. The development of 4-methylpyrazole for use in methanol and ethylene glycol poisoning is such an axample. Here, cooperation between researchers in America and Scandinavia has elucidated the human safety and pharmacokinetics (Jacobsen et al., 1988; 1989; 1990), whereas case studies in France have documented the clinical effect of the antidote in several cases of ethylene glycol poisoning (Baud et al., 1987; 1988).

Conclusion

Thus, from being a "case report-science" clinical toxicology has gradually developed into a scientific dicipline of its own with strong requirements as to the quality of research. Nevertheless, results from animal experiments and properly performed case studies will both remain important ways of getting new information. Every clinician treating a rare case of poisoning should keep this in mind as he/she has the responsibility of collecting and later present the unique features of this case, e.g. in the rare event of treating a trimethyltin poisoning (Kreyberg et al., 1991).

REFERENCES

Aarseth HP, Bredesen JE, Grynne B, Lyngdal PT, Storstein L, Wiik Larsen E. Benzodiazepine antagonist, a clinical double blind study. J Toxicol Clin Toxicol 1988, 26, 283-92.

Bismuth C, Baud F, Fournier E, Mellerio F. RO 15-1788 dans l'intoxication volontaire par les benzodiazepines. Valeur diagnostique et therapeutique. Annales de Medecine Interne 1986, 137, 363-4.

Clausen TG, Wolff J, Carl P, Theilgaard A. The effect of the benzodiazepine antagonist, flumazenil, on psychometric performance in acute ethanol intoxication in man. Eur J Clin Pharmacol 1990, 38, 233-6.

Ding GS, Liang YY. Antidotal effects of dimercaptosuccinic acid. J Appl Toxicol 1991, 11, 7-14.

Ding GS, Liang YY, Shi J, Chen W, Gu Z, Jin J. Antidotal effects of sodium dimercaptosuccinate on the intoxications from many metals (in Chinese). Chinese Medical Journal 1965, 57, 304-7.

Ekeberg Ø, Ellingsen Ø, Jacobsen D. Suicide and other causes of death in a five-year follow-up of patients treated for self-poisoning in Oslo. Acta Pshychiatr. Scand 1991, 83, 432-7.

Jacobsen D et al. A prospective study of 1212 cases of acute poisonings. General epidemiology. Hum Toxicol 1984, 3, 93-106.

Jacobsen D, McMartin KE. Methanol and ethylene glycol poisonings. Med Toxicol 1986, 1, 309-34.

Jacobsen D, Sebastian CS, Blomstrand R, McMartin KE. 4-methylpyrazole: Safety in human volunteers after single ascending doses. Alcoholism Clin Exp Res 1988, 12, 516-22.

Jacobsen D, Barron SK, Sebastian CS, Blomstrand R, McMartin KE. Non-linear kinetics of 4-methylpyrazole in healthy human subjects. Eur J Clin Pharmacol 1989, 37, 599-604.

Jacobsen D, Barron SK, Sebastian CS, McMartin KE. Effects of 4-methylpyrazole, methanol/ethylene glycol antidote, in healthy humans. J Emerg Med 1990, 8, 455-61.

Kreyberg S, Torvik A, Bjorneboe A, Wiik Larsen E, Jacobsen D.Trimethyltin poisoning. Report of a case with postmortem examination. Clin Neuropathol, in press.

Nuotto E, Palva ES, Seppala T. Naloxone-ethanol interaction in experimental and clinical situations. Acta Pharmacol Toxicol 1984, 54, 278-84.

Ritz R, Zuber M, Elsasser S, Scollo-Lavizzari G. Use of flumazenil in intoxicated patients with coma. A double-blind placebo-controlled study in ICU. Intensive Care Med 1990, 16, 242-7.

Tephly TR, McMartin KE. Methanol metabolism and toxicity. In Stegink & Filer (eds) Aspartame: physiology and biochemistry, pp. 111-140, Marcel Dekker, New York,1984.

Rygnestad T. A prospective 5-years follow-up study of self-poisoned patients. Acta Psychiatr Scand 1988, 77, 328-31.

Clinical Toxicology

(Chair: B. SANGSTER, Rijswijk)

DIAGNOSIS IN CLINICAL TOXICOLOGY

A. Jaeger, J. Kopferschmitt, Ph. Sauder, F.Flesch, C. Tournoud

Service de Réanimation Médicale et Centre Anti-Poisons - Hopital Civil - 1 place de l'Hopital - 67091 Strasbourg France

Diagnosis in clinical toxicology is based on the same principles as diagnosis in other fields of medecine. The global diagnostic approach to the poisoned patient is based on the evaluation of the history, the symptomatology, the results of the biomedical investigations and of the toxicological analytical data, and in some cases on therapeutic tests. All these data should correlate with the implicated toxin(s). The exact diagnosis should contribute to the assessment of the prognosis and indicate appropriate treatment which may include symptomatic and supportive measures, decontamination, enhancement of elimination and antidotes. However, a precise diagnosis with the exact knowledge of the toxin is rarely the first priority in an emergency situation. The early management of the poisoned patient comprises a brief examination with the evaluation of the vital signs, and supportive measures if necessary.

CASE REPORTS

The following brief case reports illustrate the different diagnostic processes. In all these cases the cause of the poisoning was initially unknown. The processes which lead to the etiologic diagnoses were initiated by history, symptoms, biomedical investigations or therapeutic tests.

Case 1

A young couple presented to the Poison control Centre because they thought they were poisoned. After dinner on Chritmas eve they had developed headache, vomiting and weakness. Their guests had the same symptoms. The physician concluded in favour of bacterial food poisoning. During the winter they had presented several episodes of headache, weakness and dizziness. The diagnoses of their doctor were either influenza or overwork. Carefull questioning revealed that the symptoms always occured when they stayed at home, were more severe in the morning when they get up, improved when they were at their workplace and were not related to a food or medication. The diagnosis of recurrent carbon monoxide poisoning was confirmed by a technical investigation performed at their home, which revealed two sources of carbon monoxide due to the faulty working of a gas water heater and of the gas heating.

Case 2

A 16-year-old boy was admitted after a self-poisoning with 2.5 g phenobarbitone. On admission, physical examination showed grade 3 coma and hypoventilation. Treatment included artificial ventilation, gastric lavage and osmotic diuresis. Twelve hours after admission the patient developed cardiovascular shock and watery diarrhoea, symptoms which could not be related to phenobarbitone toxicity. Further questioning of the parents revealed that the patient could also have ingested colchicine, a drug his father was taking for a gout. Later the patient developed typical symptoms of colchicine toxicity, especially pancytopenia, and died on the fourth day.

Case 3

During a brief stay in Strasbourg, a 39-year-old soldier in the French foreign legion, developed confusion, restlessness and convulsions. On admission, he was comatose (grade 3) and showed convulsions, hypertonia, hypoventilation and fever (39.5°C). Blood gases showed a metabolic acidosis and the EEG paroxysms of electrical activity. The convulsions improved with anticonvulsant therapy but the patient remained comatose for several weeks. Because no history was available, several diagnoses were considered: encephalitis, metabolic or degenerative encephalopathy, intoxication and post-anoxic encephalopathy. But none of these could be confirmed by numerous investigations. The diagnostic clue was provided by a friend who visited the patient for the first time two months after admission. He revealed that the patient had stomach troubles and regularly took 30-40 g bismuth nitrate daily. This had been the case for 10-20 years. At this time, bismuth was no longer freely available in France but the patient had stored the medication and had continued this "treatment" until recently. The diagnosis of bismuth encephalopathy was confirmed by the analyses performed on the biological samples which had been drawn an admission and stored. Bismuth concentration was 575 μg/l in plasma, 918 μg/l in urine and 98 μg/l in cerebrospinal fluid.

Case 4

A 41-year-old man was admitted for a bacterial pneumonia. Two hours before admission he had developed shivering, fever (39°C), dyspnoea, cough and bronchospam. Blood analysis showed hypoxemia (PaO_2=65 mmHg) and a leukocytosis of 10900/mm^3. Chest x-ray showed infiltrates in the lower lobe of the left lung. History revealed that six hours before the onset of symptoms the patient was exposed at his workplace, without protection, to metal fumes during the smelting of metals containing 90% tin.

Case 5

A 34-year-old man was admitted six hours after the ingestion of 100 mg diazepam and 2 litres of gesaprime, an herbicide containing triazine. On admission, the patient was in grade 2 coma, the blood pressure was 150/90 mmHg. Blood analysis showed an osmolar gap of 68 mMol/l and a metabolic acidosis (PH=7.27, PCO_2=2.9 kPa, bicarbonate=10 mMol/l) with an anion gap of 22 mEq/l. Methanol or ethylene glycol poisoning was suspected and the patient was treated with 4-methyl pyrazole and recovered. The initial plasma ethylene glycol concentration was 6.4 g/l. A request to the manufacturer revealed that the solvent of the herbicide contained alkylene glycol and ether polyglycols.

Case 6

A 17-year-old boy was admitted because of a coma and vomiting. He was under treatment with ofloxacine for otitis and with chloroquine because he had been in Africa 3 weeks previously. Examination showed coma (Glasgow score 5), hypertonia and vomiting. Twenty mg diazepam were administered in order to intubate the patient and perform gastric lavage. Because the patient recovered consciousness five minutes after the diazepam injection, the physician suspected chloroquine overdosage which was later confirmed by the patient's account and a plasma chloroquine concentration of 3.9 mg/l. Since diazepam has been proposed as an "antidote" in chloroquine poisoning it has been shown that these patients are rather insensitive to the central nervous effects of diazepam.

Case 7

Six hours after being sprayed with nitrobenzene during his occupation, a 39-year-old man developed weakness and cyanosis. On admission, blood analysis showed hypoxemia (PaO_2=60 mmHg), methemoglobinemia of 27% and hemolysis with a plasma hemoglobin level of 17.3 mMol/l. Treatment with methylene blue and ascorbic acid was ineffective: methemoglobinemia increased to 37.5% on the second day and decreased only after the third day. Investigation undertaken to explain the inefficiency of the antidotal treatment showed that the patient had in glucose-6-phosphate dehydrogenase deficiency (1.7 IU/g Hb: normal=10.7+/-2.5) and diaphorase deficiency (1.4 IU/g Hb: normal=2.7+/-0.5).

Case 8

A 42-year-old man consulted the Poison Control Centre because he had experienced two episodes of antabuse reaction after the ingestion of small amounts alcohol. For two weeks, during his occupation, he had handled a product used to unstick fitted carpets. According to the manufacturer the product contained only methylene chloride and methanol which could not account for the antabuse effect. Another request to the manufacturer revealed that the product contained about 17% dimethylformamide, a toxin which may induce antabuse reactions.

Case 9

A 62-year-old man was admitted to the intensive care unit because of anuria and anemia. Investigation showed intra-vascular hemolysis (plasma hemoglobin concentration = 220 mMol/l), anemia (hemoglobin = 9.8 g/l), slight methemoglobi-nemia at 3% and acute renal failure (plasma creatinine = 1030 mMol/l). Treatment included an exchange transfusion and hemodialyses (11 times). Inquiry showed that, 48 hours before admission, the patient had used, as an herbicide, a product which he had taken home from his workplace.

Toxicological analysis of the product showed that it contained basically nitrobenzene which accounted for the symptomatology.

Case 10

A 70-year-old woman, hospitalized in an internal medecine unit, suddenly developed bradycardia and shock which did not improve with cardiac pacing. On admission to the intensive care unit, examination showed cardiogenic shock with pulmonary oedema, anuria and metabolic acidosis. Because the ECG showed atrio-ventricular and intraventricular block, overdosage with an antiarrhythmic drug was suspected and the patient was treated with infusions of molar bicarbonate and isoprenaline. The features improved within one hour and the patient recovered without sequelae. In fact, the patient who had a past-medical history of hypertension, cardiac and renal failure had been treated for six days with propafenone, 600 mg daily. The plasma propafenone concentration was 1385 μg/ml in a blood sample drawn 10 hours after the last dose . Although this level did not reflect an acute overdose, it accounted for the cardiac toxicity observed in this patient with underlying cardiac disease.

Case 11

A 51-year-old woman was admitted to the intensive care unit with anuria and hemolysis. Investigations showed intravascular hemolysis (plasma hemoglobin concentration = 143 mMol/l), acute renal failure (serum creatinine = 750 μMol/l) and a hyperchloremic metabolic acidosis (PH=6.80; bicarbonate = 6 mMol/l; chlore=126 mMol/l). Treatment included infusion of bicarbonate, exchange transfusion and several hemodialyses. Before admission, the patient had been hospitalized in a surgical unit because of pancreatitis and treated with arginine chlorhydrate infusions for 15 days and penicillin infusions at a dose of 50 M daily for 8 days. Specific antipenicillin antibodies were detected in the serum by the indirect Coombs test. The final diagnoses were: hyperchloremic acidosis due to arginine chlorhydrate overdose, immunoallergic hemolysis and nephropathy due to penicillin.

Case 12

Three members of a family, a man, his wife and his son, were admitted eight hours after the onset of shivering, diarrhoea, vomiting, dyspnoea and cough. On admission two patients showed fever, diarrhoea and pulmonary oedema and one had an erythematous dermatitis. The family doctor who sent the patients to the intensive care unit suspected mercury poisoning because he had heard the patients speaking about mercury. Moreover, he had noted that the gold jewels of the patients were white-coloured and, being the son of a jeweler, he knew that mercury forms a white amalgam with gold. However, the patients denied mercury poisoning. Nevertheless, mercury poisoning was confirmed by toxicological analyses which showed plasma concentrations of 706, 878 and 420 μg/l, respectively. The exact circumstances of this unusual mercury poisoning were only reported later by one of the patients: due to a mistake the mercury of a clinical thermometer had been heated on a gas stove and they had been exposed to mercury vapor for ten hours.

DISCUSSION

1. HISTORY

Who is intoxicated, with what, how much, and when and where the poisoning occured, are the basic pieces of information required in poisoned patients. Frequently the first information concerning the poisoning circumstances is imprecise (Ellenhorn 1988). However, meticulous questioning of the patient and/or the witnesses is an essential step which may give diagnostic clues and, therefore, avoid sometimes useless investigations. For instance, poisoning involving several people is stongly suggestive of a toxin contained in food or air (case 1-12). The occurence of central nervous system disturbances or coma in a person staying in a bathroom with a gas water heater is probably related to carbon monoxide poisoning.

Epidemiological data are also helpful because they direct the first analysis towards the most common toxins (Ellenhorn 1988). Given that the epidemiology may vary between contries and is dependent on the products manufactured and on drug prescriptions, it is very useful to follow the local situation regularly. Tables 1, 2 and 3 show the causes of poisoning in 984 patients admitted to the intensive care unit during 1989. Pharmaceuticals, alcohol and drugs of abuse were the most frequent poisonings. Among the drug overdoses, about 65 per cent were due to more than one drug and bezodiazepines were involved in 91.8 % of the cases.Among the twenty most common drugs, seventeen were psychotropic drugs and ten benzodiazepines.

Table 1. Most common causes of poisoning in 984 patients admitted to the intensive care unit in 1989

	Number of cases	per cent
Pharmaceuticals	807	82.0
Alcohol	280	28.4
Drugs of abuse	40	4.1
Carbon monoxide	27	2.7
Industrial products	24	2.4
Pesticides	11	1.1
Household products	9	0.9
Mushrooms, plants	8	0.8

Table 2. Most common causes of drug overdoses in 807 patients admitted to the intensive care unit in 1989

	Number of cases	per cent
Benzodiazepines	741	91.8
Analgesics, NSAI	177	21.9
Tricyclic antidepressants	97	12.0
Neuroleptics	87	10.7
Carbamates	42	5.2
Barbiturates	21	2.6
Other psychotropic drugs	22	2.7
Cardiotropic drugs	14	1.7

Table 3. Twenty most frequent drugs involved in 807 case of acute drug overdose (frequency in per cent)

Bromazepam	19.4	Oxazepam	2.8
Lorazepam	15.0	Cyamemazine	2.8
Flunitrazepam	11.8	Ac. salicylic acid	2.8
Clorazepate-K	10.3	Amitriptyline	2.7
Paracetamol	8.6	Prazepam	2.6
Triazolam	5.6	Meprobamate	2.3
Diazepam	5.1	Febarbamate	1.7
Alprazolam	3.2	Acepromazine	1.7
Clobazam	2.9	Glafenine	1.5
Zolpidem	2.9	Haloperidol	1.5

Past medical history, underlying diseases and medications taken by the patient are the clues to diagnosis of chronic overdose, for instance with lithium, theophylline and digoxin.

If the symptomatology or the course of the poisoning is not related to the toxin which is supposed to have been ingested, it is necessary to interrogate the patient or the witnesses again (case 2-3), or to check the exact composition of the product (case 5-8-9).

2. SYMPTOMATOLOGY

Symptoms and complete physical examination may identify poisoning syndromes which provide clues to the etiology (Bismuth 1987; Ellenhorn 1988). For instance, coma, convulsions and cardiac conduction disturbances suggest a tricyclic antidepressant poisoning; coma, hypoventilation and miosis suggest narcotic overdose. Barbiturate poisoning may be suspected in a comatose patient with bullous and vesicular skin lesions. Dystonic reactions with oculo-cephalogyric crises and torsion spasms are commonly seen in children with butyrophenone poisoning. Toxic syndromes involving the cholinergic or sympathetic systems suggest specific types of toxins. An association of muscarinic and nicotinic effects is highly suggestive of organophosphate poisoning.

One of the goals of the clinician is to differentiate the symptoms which are related to a direct toxic effect from those which are complications, and to determine the mechanism of toxicity. For instance, rhabdomyolysis may be due to a toxic effect or to muscle compression during a coma. Acute renal failure is observed after poisoning with nephrotoxic substances but also after shock or intravascular hemolysis (case 9). Moreover the symptoms should be interpreted in the context of the age and the past-medical history of the patient. Severe hypoventilation may occur in patients poisoned with relatively low doses of benzodiazepines. Cardiovascular symptoms occur more rapidly and are more severe in poisonings with cardiotoxic drugs when the patient has underlying cardiac disease (case 10).

3. BIOMEDICAL INVESTIGATIONS

Although toxicological analysis is one of the main steps in diagnosis, other biomedical investigations may provide diagnostic clues (especially for toxins which are not usually included in routine screens) and are sometimes more useful in management than the measurement of the plasma drug concentrations. Moreover, biochemical disturbances may reflect the toxin's biological effects. Biomedical investigations should always be guided according by the toxin(s) involved and the symptoms.

1. Biochemical analysis

Anion gap. Metabolic acidosis may result from several causes. An anion gap (more than 12-14 mEq/l) indicates the presence in the blood of anions which are not usually measured. The most frequent cause is the accumulation of lactates resulting from cellular anoxia. Some toxins induce an abnormal anion gap because they interfere with intermediate metabolism (salicylates, ethanol) or produce accumulation of organic acid derivatives (methanol, ethylene glycol) (Jacobsen 1986).

Osmolar gap. The osmolar gap is the difference between the osmolality measured and that calculated from the plasma concentrations of sodium, glucose and urea. The most frequent cause of an osmolar gap is ethanol poisoning. The association of an osmolar gap and an anion gap is highly suggestive of a poisoning with methanol, ethylene glycol or acetone (case 5)).

Electrolyte balance. Hyperkalemia is the best criterion of severity in digitalis poisoning (Dally 1981). In chloroquine poisoning hypokalemia is well correlated with the dose ingested and with the severity of the symptomatology (Jaeger 1987). Hypocalcemia reflects the chelating effect of fluoride poisoning.

Hemolysis, methemoglobinemia are observed in poisonings by several specific toxins and may guide interrogation of the patient (case 10). The follow up of hemoglobin or methemoglobin levels is also a guide to treatment. Hemolysis may be due to a direct toxic mechanism (chlorates, arsine gas), to an immunoallergic mechanism (penicillin, case 11), to methemoglobinemia (nitrobenzene, case 9), or it may be predisposed to by an enzyme deficiency (case 7).

Hemostasis. Hemostasis disturbances may result from a direct toxic effect (anticoagulant agents, snake venoms), from hepatic failure (amatoxins) or from complications (disseminated intravascular coagulation).

Plasma cholinesterases are decreased in organophosphate poisoning. In 19 cases of organonophosphate poisoning, we noted a good correlation between the plasma cholinesterase activity and the symptoms. Moreover in five patients the evolution of the plasma cholinesterase levels well correlated with the clinical course.

2. Radiographs

A chest radiograph should be performed in all patients with coma or respiratory symptoms. Radiographs of the abdomen may indicate the presence of radiodense substances such as heavy metals or chloride solvents (Bismuth 1987; Ellenhorn 1988). In comatose patients, CAT scan of the brain may be indicated in order to eliminate a structural cause. It may also indicate cerebral post-metabolic or post-anoxic lesions in patients who present with CNS sequelae after acute poisoning. Recently, in a patient hospitalized in our unit with paraquat poisoning, a CAT scan of the chest revealed severe lesions of the lungs whereas the chest x-ray showed only minor abnormalities.

3. Electrocardiogram

ECG is the most relevant investigation in cardiotoxic drug poisonings. For instance, the association of atrioventricular and intraventricular conduction disturbances suggest a poisoning by drugs with a quinidine-like effect: class I antiarrhythmic drugs, tricyclic antidepressants, chloroquine, quinine (case 10) (Bismuth 1987). Hyperkalemia may also induce the same disturbances. Treatment includes infusion of hypertonic sodium solutions and of sympathomimetic agents such as isoprenaline or adrenaline.

4. Hemodynamic investigation

Hemodynamic studies are not routine investigations but may be helpful in the treatment of severe cardiovascular disturbances observed in poisoning with cardiotoxic drugs, calcium channel blockers or colchicine for instance (Jaeger 1990; Sauder 1983).

5. Electroencephalogram

Electroencephalogram is useful in the diagnosis and the follow up of poisoning with toxins which induce central nervous system disturbances. Specific abnormalities are noted in chloralose poisoning with myoclonic jerks. In barbiturate poisoning the EEG displays a progression of changes depending on the depth of coma and on plasma barbiturate levels. Moreover an isoelectric EEG may be observed without the presence of irreversible coma.

6. Endoscopy

Endoscopy is the preferred method for assessment of the severity of the injury in poisonings by ingestion of acids or alkali.

4. TOXICOLOGICAL ANALYSIS

Toxicological analysis may be indicated for a diagnostic, prognostic, therapeutic, medicolegal or scientific purpose.

The clinician toxicologist may be confronted by three situations:

- poisoning is certain, the toxin(s) is known according to the account of the patient or the witnesses, the symptoms are related to the toxin(s). In this case toxicological analysis is not essential if it has no prognostic, therapeutic or medicolegal implications.
- the poisoning is certain, the toxin(s) is known but the symptoms are not related to the suspected toxin(s) or to the dose. A toxicological analysis is indicated in order to detect other toxins which may have been ingested. A negative result should direct the investigation towards a non-toxic cause.
- the history and the symptomatology suggest a toxic cause but the toxin(s) is unknown. Only a toxicological analysis may confirm poisoning in this case.

The usefulness of systematic toxicology screens in overdoses is not established. In several studies, drugs judged responsible for the overdoses were only identified in 26 to 86 % of the cases. In 14 to 28 % of the cases others drug were identified (Hepler 1986; Rygnestad 1984). In a personal study of 205 cases of acute poisoning with five toxins which are frequently involved, the ingested drug according to the history was identified in 62.5 to 100 % of the cases (table 4). Antidepressants, paracetamol, salicylates were indentified in 2.4% to 6.3% of the patients who did not recognize they had ingested these drugs. These false positive results were mostly due to chronic treatment. Benzodiazepines and ethanol were identified in 28.4 and 30.7 % of the cases although the patients denied the ingestion of these toxins.

Toxicology screens give mostly qualitative or semi-quantitative results for drugs which are frequently involved in poisoning. But they do not include most of the toxicons for which the analysis is essential for a prognostic or a therapeutic purpose.

Table 4. Toxicology screen in 205 cases of acute poisoning with benzodiazepines, antidepressants, paracetamol, salicylates or ethanol.

Drug	Acute ingestion		No acute ingestion	
	Number of cases	Positive analysis(%)	Number of cases	Positive analysis(%)
Benzodiazepines	120	84.2	95	28.4
Antidepressants	8	87.5	207	6.3
Paracetamol	10	100	205	2.4
Salicylates	8	62.5	207	2.4
Ethanol	88	88.6	127	30.7

A quantitative analysis should be ordered to predict future toxic effects (salicylates, paracetamol, paraquat), to improve or confirm a diagnosis (barbiturates, carbamates, carbon monoxide) or to guide treatment with hemodialysis (methanol, ethylene glycol, lithium), hemoperfusion (barbiturates, carbamates, theophylline) or antidotes (paracetamol, digitalis) (Ellenhorn 1988; Proudfoot 1986; Rumack 1975). Toxicokinetic studies are sometimes indicated in order to determine the indication of specific treatments or to confirm their efficiency (Jaeger 1990).

Analysis of the product which has been ingested or of the air to which the patient was exposed may also be performed in order to retrospectively confirm the diagnosis (case 1-9).

5. THERAPEUTIC TESTS

In some poisonings the administration of an antidote has a therapeutic and diagnostic value. Moreover, the history can be completed if the patient regains consciousness after the administration of the antidote. The typical antidotes which may be used for therapeutic tests are naloxone for narcotic or opioid analgesics, glucose for hypoglycemics, flumazenil for benzodiazepines. We studied the effect of flumazenil administration in 162 patients with consciousness disturbances due to benzodiazepine poisoning. 31.5 % of the patients had only ingested benzodiazepines and 68.5% had also ingested other psychotropic drugs or alcohol. 94.4 % of the patients regained consciousness after flumazenil injection and, therefore, gastric lavage could be performed without intubation in 104 cases and 8 out of 13 patients previously intubated could be extubated immediately.

A lack of efficacy of the antidote does not exclude the diagnosis either because other psychotropic drugs may be associated with opiates, benzodiazepines, or because brain damage may have occured during hypoglycemia. A lack of efficacy may also indicate underlying disease. An incomplete recovery may be observed after treatment of digitalis overdose with digoxin antibodies in patients with chronic heart disease. A lack of effect of antidotes for methemoglobinemia indicated an enzyme deficiency in one of our cases.

CONCLUSION

Diagnosis in clinical toxicology is based on history, symptomatology, biomedical investigations and therapeutic tests. The clinician toxicologist has to collect the data which should correlate with the toxin(s). Diagnosis is never or rarely a matter of chance. Biomedical investigations and toxicological analysis especially should be guided by history and symptoms. The final goal is to establish a precise diagnosis so that treatment is appropriate to the patient and the toxin. Precise diagnosis is also mandatory for adequate epidemiological studies, for the assessment of prognostic criteria and dose-symptomatology relations, and for the evaluation of the efficacy of treatments.

REFERENCES

Bismuth C (1987) Toxicologie Clinique. Flammarion, Paris

Dally S, Alperovitch A, Lagier G, Bismuth C, Fournier E (1981) Facteurs pronostiques de l'intoxication digitalique aigue. Nouv Presse Med 10: 2257-2260

Ellenhorn MJ, Barceloux DJ (1988) Medical Toxicology. Elsevier New York Amsterdam London

Hepler BR, Sutheimer CA, Sunshine I Role of the toxicology laboratory in the treatment of acute poisoning (1986) Med Toxicol 1: 61-75

Jacobsen D, McMartin KE (1986) Methanol and ethylene glycol poisonings: mechanism of toxicity, clinical course, diagnosis and treatment. Med Toxicol 1: 309-334

Jaeger A, Sauder Ph, Kopferschmitt J, Flesch F (1987) Hypokaliemie au cours des intoxications par la chloroquine. Presse Med 33: 1658-1659

Jaeger A, Sauder Ph, Kopferschmitt J, Dahlet M (1990) Toxicokinetics in clinical toxicology. Acta Clin Belg 45: 1-12 (suppl. 13)

Jaeger A, Sauder Ph, Bianchetti G, Kopferschmitt J, Dahlet M (1990) Intoxications aigues par le diltiazem. Etude cinétique et hémodynamique. J Tox Clin Exp 10:261-270

Proudfoot AT, Stewart MS, Levitt T, Widdop B (1986) Paraquat poisoning. significance of plasma-paraquat concentrations. Lancet 2: 330-332

Rumack BM, Matthew H (1975) Acetaminophen poisoning and toxicity. Pediatrics 55: 871-876

Rygnestad T, Berg KJ (1984) Evaluation of benefits of drug analysis in the routine clinical management of acute self poisoning. B Med J 289: 426-428
Sauder Ph, Kopferschmitt J, Jaeger A, Mantz JM (1983) Hemodynamic study in eight cases of colchicine poisoning. Hum Toxicol 2: 165-167

HAEMOPERFUSION, INDICATIONS AND SIDE-EFFECTS

J.H. Rommes, MD
Department of Critical Care/Pulmonary Care Medicine
University Hospital
Groningen, the Netherlands

INTRODUCTION

Acute intoxications pose a significant public health problem in most countries. In the Netherlands approximately 13.300 patients were admitted to the hospital due to accidental or intentional overdose and poisoning in 1989.
Although overdosed patients are often critically ill, current methods of intensive care keep the mortality rate relatively low; less than one per cent of those admitted to the hospital die (Vale 1977; Garella and Lorch 1980; Cutler et al 1987).
The management of intoxicated patients is based on prevention of further absorption, enhancement of elimination, in rare instances administration of antagonists, and supportive care.
A number of methods are used to enhance elimination of drugs or toxic compounds from the body such as forced diuresis with pH control, haemodialysis, haemoperfusion, plasma exchange,peritoneal dialysis and haemofiltration. The clinical importance of an effective method to remove drugs or pharmacologically or toxicologically active metabolites from the body is illustrated by the fact that the mortality rate from severely intoxicated, deeply comatose patients varies from 13 to 35 per cent (Afifi et al 1971; Arieff et al 1973; Hampel et al 1979) despite intensive supportive therapy. Furthermore, complications of toxic ingestion are related to the degree and duration of coma (Moeschlin 1971, Greenblatt et al 1979, Gelfland and Winchester 1980).

INDICATIONS FOR HAEMOPERFUSION.

The most logical therapy for acute intoxications is to remove the toxin. Haemoperfusion has a both in vitro and in vivo proven ability to accelerate the removal of a wide range of intoxicating substances (Lorch and Garella 1979; Gelfland et al 1977; Yatzidis et al 1965, Winchester et al 1977; de Groot 1982). Clinically, in retrospective studies Vale et al (1975) have demonstrated that haemoperfusion over coated charcoal significantly improved survival of patients with severe intoxications. Hampel et al (1978) found in a prospective analysis a mortality rate of 20% in severely intoxicated patients not treated with haemoperfusion; in the group patients having similarly severe intoxications who received haemoperfusion mortality was 7%. De Groot (1982) concluded based on the results of an extensive pharmacokinetic investigation: "there is a clear and convincing role for haemoperfusion in the treatment of a selected group of severely poisoned patients ". However, the question "does haemoperfusion markedly improve prognosis or decrease mortality in exogenous poisonings" remains to be answered in a prospective randomised study.

The clinical setting and some of the toxic substances for which there is an indication for the use of haemoperfusion have been delineated during a meeting organized by the advisory subgroup on toxicology of the European Medical Research Council in 1984 (de Broe et al 1984). During that meeting it was concluded that indications for haemoperfusion in the management of the poisoned patient are:

- an established intoxication causing a severe disturbance of one or more of the vital functions with a drug having a small distribution volume and displaying a clear relation between blood levels and toxic effects.
- a deterioration of the clinical state of an intoxicated patient with initially less severe symptoms, despite optimal conservative treatment.
- an intoxication with a drug known to be metabolized to a more toxic one.
- an intoxication with a drug known to produce delayed toxicity.

drug/poison	preferred adsorbent regarding extraction	reference
Amitryptilyne	C	24
Barbiturates	R/C	12
Camphor	C	20
Carbamazepine	C/R	24
Carbon tetrachloride	C	12
Carbromal	R	12
Digoxine	C/R	12, 24
Dimethoate	?	
Diphenylhydantoine	C/R	24
Ethchlorvynol	R	12, 24
Glutethimidine	R	12, 24
Flecainide	R	13
Meprobamate	C/R	
Methaqualone	C/R	12, 24
Methotrexate	C	24
Methylphenobarbital	C/R	12, 24
Nortryptilyne	C	12, 24
Paracetamol	C	21
Paraquat	C	24
Thallium	R/C	
Theophylline	C/R	24

Table 1 A number of drugs and poisons eliminated during haemoperfusion over activated charcoal (C) or resin (R).

BIOCOMPATIBILITY OF HAEMOPERFUSION

In order to characterize biomaterials the term "biocompatibility" has been introduced as a quality concept. This neologism is derived from the Greek "βιοσ" and compatibility derived from the latin "compatibilis".

The term "biocompatibility" is usually considered to imply that the clinical use of such materials will not create effects that endanger the lives of patients.
With an increased use of haemoperfusion as an effective drug-elimination technique a number of side effects of haemoperfusion has been reported: hypotension, hypocalcemia, reduction in platelet and white blood cell counts, an increased infection rate and disturbances of the haemostasis resulting in an increased bleeding tendency (Pond 1984). The haemoperfusion related complications determine the degree of biocompatibility of this procedure. Because haemoperfusion is used in the management of critically ill, poisoned patients it is obvious that the clinician must be aware of the severity and extent of these side-effects. The clinically most relevant potential hazards of haemoperfusion are infections, bleeding and hypotension.

Infection

Obviously, several factors such as the severity of illness, hypoventilation causing atelectasis, and aspiration contribute to an increased risk of infection in critically ill poisoned patients who are treated with haemoperfusion. However, several investigators suggest that haemoperfusion itself induces an additional risk of infection as a consequence of a decrease of the concentration of immunoglobulins and complement factors due to adsorption to the perfusion column (Hennemann et al 1977, Bismuth 1980). Another factor possibly contributing to an increased risk of infection could be the occurrence of a haemoperfusion related leukopenia due to trapping in the perfusion column (Kolthammer et al 1976, Sangster et al 1981). In a study in severely poisoned patients it was demonstrated that haemoperfusion induces a slight (i.e. 8-15%) decrease of the immunoglobulin levels (IgA, IgM and IgG), total haemolytic complement (CH50) and complement factors C3 and C4 (Rommes et al 1983). This decrease reflects haemodilution induced by haemoperfusion. Furthermore a transient reduction of the white cell count involving particularly granulocytes was observed. The decrease of immunoglobulin and complement concentrations and granulocyte count do not nearly reach the levels associated with an increased risk of infection. It was concluded that regarding the host defense related to immunoglobulin concentrations, complement system and white cell counts, haemoperfusion does not induce an increased susceptibility to bacterial infections (Rommes et al 1983). Further studies are required to evaluate the influence of haemoperfusion on the function of the granulocytes. Granulocytes show a markedly depressed phagocytic activity after artificial surface contact (Vanholder 1989).
Most of the basic and clinical information on blood-artificial interactions has been obtained from studies of haemodialyzers. These interactions include complement activation via the alternative pathway, leading to production of the anafylatoxins C3a and C5a. Generation of the potent biological effector C5a is responsible for granulocyte activation, aggregation and subsequently sequestration of these cells in the pulmonary microvasculature during haemodialysis. Particularly cellulosic haemodialysis membranes exhibit a pronounced complement-activating potential (Henderson 1987). The interaction between

blood and the haemoperfusion column particularly the column containing cellulose coated activated charcoal induces complement activation in view of the detection of C3d as a biological marker of complement activation (Rommes et al 1983).

In standardized animal experiments haemoperfusion was performed to investigate the nature of the haemoperfusion related leukopenia (Rommes et al 1983; 1986). The differences in white cell and granulocyte counts in the inlet and outlet indicate trapping of these cells in the perfusion column. Furthermore, a right ventricle-aorta gradient of the granulocyte count indicating sequestration of granulocytes in the pulmonary capillaries, was observed during the initial phase of haemoperfusion. Hence, the reduction of the white cell count is caused by entrapment of white cells not only in the perfusion column but also in the pulmonary microvasculature. From these observations it was concluded that haemoperfusion over cellulose coated activated charcoal and to a lesser extent resin, induces complement activation and complement mediated granulocyte sequestration in the pulmonary microvasculature (Rommes et al 1986).

Cardiopulmonary changes

What are the clinical implications of haemoperfusion induced complement activation? Craddock et al (1977) postulated in their original hypothesis that pulmonary leukosequestration results in the hypoxaemia commonly observed during haemodialysis. Other data suggest that this hypoxaemia results from hypoventilation induced by an alteration of the respiratory quotient as consequence of carbon dioxide loss across the dialysis membrane (Aurigemma et al 1977, Henderson et al 1987). Haemoperfusion associated hypotension is thought to result primarily from hypovolemia. Other factors possibly contributing to hypotension are hypocalcemia and catecholamin depletion due to adsorption to the perfusion column (Sorkin et al 1982, Pond 1984).

In standardized animal experiments haemoperfusion was performed to clarify the cardiopulmonary effects of haemoperfusion (Rommes et al 1986). General anaesthesia, muscle paralysis and mechanical ventilation excluded the possibility of hypoventilation. Arterial oxygen tension decreased during haemoperfusion; however no differences with the control group were observed (Rommes et al 1986). These data indicate that complement mediated pulmonary leukostasis does not produce hypoxaemia. However, it should be noted that the extent of pulmonary leukostasis during haemoperfusion appears to be limited. Due to sequestration of granulocytes in the perfusion column the granulocyte count in the right ventricle and thus the number of cells involved in pulmonary leukostasis is relatively low.

Analysis of the haemodynamic mechanisms involved in the profound drop of the blood pressure immediately after the start of charcoal haemoperfusion reveals the following: a decrease of stroke volume and heart rate both contributing to a decrease of the cardiac output and a lack of compensatory increase of the systemic vascular resistance.

Resin perfusion induces similar, although less pronounced, haemodynamic changes.

It is obvious that these cardiovascular changes are at least partly a consequence of hypovolaemia. However, the haemodynamic changes observed during haemoperfusion are in contrast with the, according to most textbooks, "normal" responses to hypovolaemic hypotension, i.e. tachycardia and peripheral vasoconstriction. Recent observations in patients and experiments in animals indicate that bradycardia and vasodilatation may occur in hypovolemic shock (Secher et al 1985, Sanders et al 1989). Central depression of autonomic mechanisms by release of endorphins or paradoxical activation of the ventricular mechanoreceptors have been proposed as mechanisms for this sympathicoinhibition (Sanders et al 1989). Haemoperfusion over cellulose coated charcoal induces a decrease of the serum calcium concentration (Bentley et al 1979). However, the level of serum ionized calcium did not reach levels associated with cardiovascular complications (Rommes et al 1986).

The simultaneous occurrence and transient character of the cardiovascular changes, changes in white cell and platelet counts and complement activation may indicate an interrelationship. The question arose whether these phenomena are specific for haemoperfusion or could be the result of blood-artificial surface interaction. During haemodialysis similar haemodynamic and haematological changes are observed (Rommes et al 1987). The fact that during haemodialysis serum calcium concentration did not change and removal of catecholamines probably was prevented, indicates that these cardiovascular changes are not related to hypocalcaemia or catecholamine depletion. Simultaneously with the cardiovascular changes a decrease of the platelet count due to sequestration in the dialyzer and a reduction of the white cell count due to complement mediated pulmonary leukostasis was demonstrated. No significant changes of the arterial oxygen tension were observed.

Summarizing the results of these experiments it is concluded that the blood-artificial surface interaction occurring during haemoperfusion over cellulose coated activated charcoal and resin, and cuprophane haemodialysis induces complement mediated pulmonary leukostasis without decrease of the arterial oxygen tension.

The haemodynamic changes occurring in the first 30 minutes of both charcoal haemoperfusion and cuprophane haemodialysis are characterized by a drop of stroke volume and heart rate resulting in a decrease of the cardiac output without a compensatory increase of the systemic vascular resistance. Resin perfusion induces comparable although less pronounced changes (Rommes et al 1983, 1986). The striking similarity of the haemodynamic changes during charcoal haemoperfusion and cuprophane haemodialysis makes it tempting to speculate on the role of the pronounced complement-activating potential of cellulose coated charcoal and cuprophane in the pathogenesis of these haemodynamic changes. Complement activation, generation of anafylatoxins, granulocyte and monocyte activation induce release of inflammatory mediators such as elastase, interleukin-1, leukotrienes and prostaglandins (Dinarello 1988).

The vasoactive properties of leukotrienes and prostaglandins may contribute to the cardiovascular symptoms, particularly the lack of increase or even decrease of the systemic vascular resistance, during haemoperfusion and haemodialysis.

Further studies should be performed to evaluate the role of complement activation and it sequelae in the pathophysiology of the cardiovascular phenomena occurring during the initial phase of haemoperfusion and haemodialysis.

Bleeding

The increased bleeding tendency provoked by haemoperfusion has been attributed to platelet depletion due to sequestration in the perfusion column and to the effect of heparin. Furthermore haemoperfusion may induce a platelet function disturbance (Sangster et al 1981). Platelet sequestration in the perfusion column appears to be a transient phenomenon occurring during the initial 30-40 minutes of perfusion followed by an slight increase of the platelet count (Rommes et al 1983, 1986). In vitro perfusion experiments show that this increase is for a minor part caused by release of platelets from the column. Therefore, the increase of the platelet count has to be attributed to release from bone marrow and spleen. Platelet aggregation, measured as the functional response on collagen and ADP stimulation, appears to be slightly affected by haemoperfusion over charcoal and resin. The slight increase of the plasma ß-thromboglobulin (ß-TG) level, a marker of the secretory α-granules, after perfusion indicates that haemoperfusion induces platelet secretion only to a limited extent. Furthermore, it appears that after haemoperfusion the serotonin, ß-TG, ADP and ATP contents of the circulating platelets are intact (Rommes et al 1991). From these data it was concluded that haemoperfusion over cellulose coated charcoal and resin induces a platelet function disturbance which can not be attributed to a storage pool deficiency as has been suggested by several authors (Weston et al 1977; Sangster et al 1981).
Furthermore it was concluded that platelet sequestration in the column cannot be attributed to activation and aggregation alone but that other mechanisms such as adhesion to the adsorbing surface and mechanical trapping are involved in this phenomenon (Rommes et al 1991).
The cause of the haemoperfusion related platelet function disturbance is thus at present unknown. A possibility is that interaction between red cells and the column resulting in release of ADP from red cells (Salzmann 1977), induces platelet activation followed by a state of refractoriness of these platelets (Peerschke 1985). Further studies are required to clarify the cause of haemoperfusion related platelet function disturbance. However, considering the slight extent of inhibition of platelet aggregation, its clinical importance is probably limited. Which factors are responsible for the transient character of the proces of platelet sequestration? The relatively functional integrity of platelets after perfusion suggests that the transient character of platelet sequestration can not be attributed to a haemoperfusion induced platelet function disorder (chapter 7). Interaction of blood cells with artificial surfaces is preceded by adsorbtion of a film of plasma proteins.
Fibrinogen is an important component of this protein layer and plays probably an essential role in platelet adhaesion and aggregation (Salzman et al 1987). Activated platelets (i.e. platelets stimulated by for example ADP) expose their glycoprotein membrane receptors to which fibrinogen binds with

high affinity. Recent data suggest that fibrinogen adsorbed to an artificial surface may change its confirmation to a form that is unrecognizable to platelet receptors (Salzmann 1987). Another possibility is that further adsorption of plasma proteins prevents direct contact between platelets and the adsorbing content of the column and thus adhesion of platelets to the artificial surface. Preliminary data indicate that priming of the haemoperfusion column with an albumin solution, thus preventing platelet-artificial surface contact can reduce the problem of platelet depletion (Chang 1984). It is evident that further studies are required to clarify the transient character of platelet sequestration. Insight in this phenomenon may provide the possibility to improve the biocompatibility of haemoperfusion columns and artificial surfaces in general.

An effective method to preserve platelet count during haemoperfusion is administration of a platelet activation blocking drug such as prostacyclin (PGI_2). Previous studies have shown that PGI_2 infusion at a rate of 16 $ng.kg^{-1}.min^{-1}$ during haemoperfusion prevents platelet loss and platelet activation (Langley et al 1979, Gimson et al 1980). The present study shows that prostacyclin infusion at a much lower rate (1 $nanogram.ml^{-1}.min^{-1}$ comparable with 3 $ng.kg^{-1}.min^{-1}$ during in vivo perfusion) into the proximal line of the column prevents platelet sequestration and secretion. Infusion at a lower rate probably will reduce the side effects of PGI_2 (hypotension due to vasodilatation, tachycardia and bradycardia). It is interesting to speculate on the mechanism of this prostacyclin related platelet preservation.

Assuming that platelet adhesion rather than platelet aggregation is the principal mechanism involved in the proces of platelet sequestration during haemoperfusion, the question arises if prostacyclin prevents not only platelet activation but also platelet adhesion. Recent data in fact indicate that prostacyclin may inhibit platelet adhesion to artificial surfaces (de Groot and Sixma; unpublished data).

Reduction of the flow during the first 40 minutes of in vitro perfusion results in improved platelet count (Rommes et al 1991). If this improved platelet preservation at lower flow rates can be attributed to a reduction of shear stress related platelet activation or to a less intense blood artificial surface interaction remains to be determined.

Hence, the clinician who worries about the risk of bleeding in the poisoned patient requiring haemoperfusion has at the moment two possibilities to preserve an improved platelet count. Firstly, reducing the flow (100 $ml.min^{-1}$ instead of 300 $ml.min^{-1}$) during the initial 45 minutes of perfusion with as disadvantage of a reduced clearance of the drug cq poison. Secondly, definitely more effective the administration of prostacyclin. The hazards of prostacyclin (hypotension and flushing) probably can be reduced by infusion in the inflow line of the perfusion column.

CONCLUSIONS

It is evident that regarding two of the three discussed biocompatibility problems resin haemoperfusion should be preferred over perfusion over cellulose coated activated charcoal. Particularly the sequelae of haemoperfusion related complement-activation requires further investigation. Studies of artificial kidney membranes have provided accumulating evidence for the concept of "whole body inflammatory reaction" provoked by blood(cell)-artificial surface interaction and mediated at least partly by the complement system (Henderson 1987, Dinarello 1985). Whether a number of the phenomena described in this thesis can be attributed to such a systemic response remains subject for further study. If so, much of the knowledge of biocompatibility of dialysis membranes may be translated to haemoperfusion columns and vice versa, ultimately resulting in an optimal biocompatibility.

REFERENCES

1. Affifi AA, Sachs ST, Liu VY, Weil MH, Shubin H. Accumulative prognostic index for patients with barbiturate, gluthemidine and meprobamate intoxication. N Eng J Med 1971;285:1497-502.
2. Arieff AI, Friedman EA. Coma following nonnarcotic drug overdosage: management of 208 adult patients. Am J Med Sci 1973;266:405-26.
3. Aurigemma NM, Feldman NT, Gottlieb M, Ingram RH, Lazarus JM, Lowrie EG. Arterial oxygenation during hemodialysis. N Eng J Med 1977;297:871-3.
4. Bismuth C, Scherrmann JM. Hémoperfusion sur charbon active enrobé. Med Legale Toxicol 1980;23:23-36.
5. Bentley C, Kjellstrand CM. The treatment of severe drug intoxication with charcoal haemoperfusion in series with hemodialysis. J Dialysis 1979;4:337-48.
6. de Broe ME, Bismuth C, de Groot et al. Haemoperfusion: a useful therapy for a severely poisoned patient? Human Toxicol 1986;5:11-5.
7. Chang TMS. Coated charcoal Haemoperfusion. Lifi Supp Sys 1982;2:99-106
8. Craddock PR. Fehr J. Brigham KL, Kronenberg PJ, Jacobs HS Complement and leucocyte-mediated pulmonary dysfunction in haemodialysis. N Engl J Med 1977;296:769-74
9. Cutler RE, Forland SC, Hammond PG, Evans JR: Extracorporeal removal of drugs and poisons by haemodialysis and haemoperfusion. Ann Rev Pharmacol Toxicol 1987;27:169-91.
10. Dinarello CA, Koch KM, Shaldon S. Interleukin-1 and its relevance in patients treated with haemodialysis. Kidney Int 1988 33(S):S21-S6.
11. Garella S, Lorch JA. Hemoperfusion for acute intoxications: Con. Clin Toxicol 1980;17(4):515-27.
12. Gelfland MC, Winchester JF. Hemoperfusion in drug overdosage: a technique when conservative management is not sufficient. Clin Toxicol 1980;17(4):583-602.
13. Götz D, Pohle S, Barckow D. Primary and secondary detoxification in severe flecainide intoxication. Int Care Med 1991;3:181-4
14. Greenblatt DJ, Allen MD, Harmatz JS, Noel BJ, Shader RI. Overdosage with pentopbarbital and secobarbital: assessment of factors related to outcome. J Clin Pharmacol 1979;19:758-68.

15. Groot de G: Haemoperfusion in clinical toxicology; a pharmacokinetic evaluation. Thesis 1982, Utrecht, The Netherlands.
16. Hampel G, Wiseman H, Widdorp B. Acute poisoning due to hypnotics: the role of haemoperfusion in clinical perspective. Vet Hum Toxicol 1979;21(S):4-6.
17. Henderson LW, Chenoweth D. Biocompatibility of artificial organs: an overview. Blood Purification 1987;5:100-11
18. Hennemann H, Richter IE, Brunswig D, Rockel A, Kult J, Gattenlohner W, Heidland A. Hamoperfusion-pro und kontra. Klin Wochenschr 1977;55:53-7.
19. Kolthammer JC, Watson PA Lang SA, Fennimore J. The safety assessment in the dog of a charcoal haemoperfusion column. Clin Sci Mol Med 1976:51;515-24.
20. Köppel C, Martens F, Schirop Th, Ibe K. Hemoperfusion in acute camphor poisening. Int Care Med 1988;14:431-3
21. Langley PG, Hughes RD, Ton HY, Silk DBA, Williams R. The use of prostacyclin, PGI_2, in the prevention of platelet aggregation during charcoal haemoperfusion. Int J Artif Organs 1979;2:207-10.
22. Moeschlin. Clinical features of barbiturate poisoning. In: Matthew H. Acute barbiturate poisoning. Excerpta Med, Amsterdam 1971;117-28.
23. Peerschke EIB. Ca^{+2} mobilization and fibrinogen binding of platelets refractory to adenosine diphosphate stimulation. J Lab Clin Med 1985;106:111-22.
24. Pond S. Diuresis, dialysis and hemoperfusion. Emerg Med Clin N Am 1984;2:29-45.
25. Rommes JH, Sangster B, Gmelig Meyling FHJ, Daha MR, Heyst ANP van: The influence of hemoperfusion on white cells, immunoglubulin concentrations and the complement system in the treatment of intoxications. Arch Toxicol 1983;52:149-56
26. Rommes JH, Sangster B, Borst C, Daha MR, Heyst ANP van: The influence of hemoperfusion on leucocytes, thrombocytes, complement, circulation and arterial oxygenation in dogs. Veterinary and Human Toxicology 1983;5:7-10
27. Rommes JH, Sangster B, Berrens L, Borst C, Heyst ANP van. Biocompatibility of hemoperfusion. Arch Toxicol 1986:58;187-95.
28. Rommes JH, Sangster B, Borst C, Daha MR, Heyst ANP van. Biocompatibility of haemodialysis using cuprophane membranes. Vet Hum Toxicol 1987;29:293-299.
29. Rommes JH, IJsseldijk M, Sangster B, Sixma JJ: The influence of haemoperfusion on electrolytes, platelet count and platelet function. Arch Toxicol: accepted for publication
30. Rommes JH, IJsseldijk M, Sangster B, Sixma JJ: Platelet count and function during haemoperfusion. Arch Toxicol: submitted for publication
31. Salzman EW, Lindon J, Brier D, Merrill EW. Surface-induced platelet adhesion, aggregation and release. Behavior of blood and its components at interfaces. L.Vroman, EF Leonard,eds. Ann NY Acad Sci 1977;283:114-21.
32. Salzman EW, Lindon J, McManama. Ware JA. Role of fibrinigen in activation of platlets by artificial surfaces. Ann NY Acad Sci 1987;516:184-96.

33. Sanders JS, Ferguson DW. Profound sympathicoinhibition complicating hypovolemia in humans. Ann Int Med 1989;111:439-41

34. Sangster B, Heyst ANP van, Sixma JJ. The influence of haemoperfusion on haemostasis and cellular constituents of the blood in the treatment of intoxications. Arch Toxicol 1981;47:269-78.

35. Secher NH, Bie P. Bradycardia during reversible haemorrhagic shock-a forgotten observation? Clin Physiol 1985;5:315-23.

36. Sorkin HL, Gunsalus E, Clark JE. Removal of sympathicomimetic amines during hemoperfusion across resin and charcoal columns. Trans Am Soc Artif Intern Organs 1982;28:396-9.

37. Vale JA. The epidemiology of acute poisoning. Acta Pharmacol Toxicol 1977;41(s):443.

38. Vanholder R, Ringoir S. Bioincompatibility: an overview. Int J Art Org 1989;12:356-65.

ACCIDENTAL POISONING WITH HYDROCARBON PRODUCTS IN CHILDREN DURING SIX MONTHS 1989.

B Lundholm, C Karlson-Stiber

Swedish Poison Information Centre, Box 60500, S-104 01 Stockholm, Sweden

INTRODUCTION

Hydrocarbon products represent a major risk of intoxication to children all over the world, and may cause severe symptoms and even death. It is therefore important to identify special risk factors to be able to take adequate preventive measures.

The following example is a typical case of hydrocarbon poisoning in childhood and illustrates circumstances frequently occurring and possible course of illness:

A little boy is playing in the garden. The parents had a barbecue the day before, and the bottle with fire lighting fuel was left behind, just for once. The child happens to find the bottle, opens it and tries to taste a little. But when he swallows, the very thin and very liquid chemical goes down the wrong way. The boy starts coughing intensely.

The parents get anxious and phone to the Poison Information Centre to ask if this might be dangerous. The hacking cough from the child, heard in the background, sounds very characteristic. During the telephone call the child starts to get pale and drowsy, and immediate transport to the hospital is necessary.

The child is admitted to the intensive care unit. He has difficulties in breathing and the oxygenation is impaired. Treatment in respirator is necessary for twenty-four hours. A pulmonary infiltrate is seen on the X-ray, and he developes a high fever. A typical chemical pneumonia.

Petroleum distillates - especially fire lighting fuels, lamp oils and white spirits - are very liquid because of low viscosity and low surface tension. When swallowed (or vomited) they are easily aspirated and spread out over the mucous membranes of the lungs. These physical properties explain why already a very little amount can give symptoms of intoxication. In the lungs an inflammatory reaction arise and together with the reduced surface tension it leads to impaired oxygenation. This chemical pneumonia can be very serious, and even fatal.

In 1989 the Swedish Poison Information Centre received 92 case records from the hospitals regarding children intoxicated with petroleum distillates. Chemical pneumonia was verified by radiographs in at least 33 cases. Severe symptoms developed in 7 children. Two of those children required treatment with respirator, and another child had transient respiratory arrest and convulsions.

The real number of children treated in hospital due to poisoning with hydrocarbon products during 1989 was probably higher, because all cases are not reported to the Poison Centre.

Thus, household products containing hydrocarbons represent a major risk of severe illness. Furthermore they are frequently involved in poisoning incidents during childhood. In order to reduce the number of such accidents the National Board for Consumer Policy in Sweden 1988 issued guidelines regarding child resistant packaging.

With the intention to evaluate the effect of these guidelines, and to identify contributory factors to poisoning incidents with petroleum products in childhood, the Swedish Poison Information Centre has carried out a study on this topic. This study was granted by the National Board for Consumer Policy.

METHODS AND RESULTS

Inquiries to the Poison Centre about Hydrocarbon Products

During six months, April to September, the Poison Centre was consulted in 676 cases of poisoning accidents with petroleum distillates in children. This indicates a considerable increase (about a doubling) since 1981 when a similar study was done.

The products most frequently involved were fire lighting fuels/ lamp oils (249 cases) and white spirits (179 cases). Specially the first group had increased considerably since 1981. This is partly depending on the introduction of oil lamps in our country, but possibly also because of a more widespread custom to barbecue. (The fuel and the oil are joint together in writing, because it is the same product in practice).

The inquiries to the Poison Centre are from the general public as well as from the hospitals.

In about 2/3 of the consultations from general public no initial symptoms had occurred. These children had probably not aspirated and could stay at home with extra supervision.

In about 1/3 of the incidents immediate symptoms were observed, mostly coughing and sometimes vomiting or drowsiness. These children had obviously aspirated. It is well known that in many of these cases the symptoms will be mild or moderate, but a few children will develope severe illness. However, in this early stage it is impossible to predict the progress of the intoxication. Consequently all children with initial symptoms were recommended hospital care.

The majority of the children were one or two years old with a predominance of boys (as in most other studies with child accidents).

A Study of 99 Incidents with the Original Bottle

To evaluate the new guidelines regarding child resistant packaging, parents to children, who had tasted the products directly from the original bottle, were interviewed by telephone about the closure of the bottle and the location of the bottle, among other things. It was possible to reach 99 of these parents.

In 48 of the 99 cases the incidence occurred when the bottle had been left behind at the place for its use (e.g. after barbequeing) and the incident took place hours or even days later. 25 children got hold of the bottle while the parents were using the product. Only in 12 incidents the product was stored in its ordinary place.

The interviews with the parents confirmed something which is well known to us - but maybe not to the parents - that a major part of poisonings occur during the use of the products when precautions are easily forgotten, or even to a greater extent, after the use is finished and the parents forget to place the bottle safely.

Regarding the packaging, it appeared that many people still had old bottles of these products in their homes, because 76 of the 99 incidents occurred with the old-fashioned packaging, easy to open. So it seemed to be a little too early to evaluate the effects of the new guidelines.

Only 23 of the involved products were equipped with the new child resistant closure, of which 6 were already opened before the child found it. Nevertheless 17 children managed to open the closed childproof cap, and sometimes surprisingly fast. Furthermore most of these children were less than 2 years old.

(In addition, during the same time period, the Poison Centre was informed about five more children who had opened other products equipped with child resistant closures.)

Consequently the child resistant packaging must never be regarded as absolutely safe, but it may give the parent a better chance to stop an accident in time.

Hydrocarbon Products in Other Containers

Let us hope that a more widespread use of child resistant closures in a near future will contribute to a decreased number of accidents with petroleum distillates in the original bottles. However, no kind of childproof caps can prevent the large number of accidents occurring when products have been transferred to other containers to be used.

Regarding lamp oils/fire lighting fuels, about the same number of incidents occurred when the liquid had been poured into an oil lamp (75 cases) as it did when the product still was in a bottle (81 cases). In this connection it is important to remember that severe poisonings have occurred even when children just have sucked the wick of the lamp.

Regarding white spirits, much more incidents occurred when the liquid had been poured into a tin with paint brushes (104 cases) than with the same product in a bottle (18 cases).

In the light of this experience, it is obvious that child resistant closures can not prevent all kinds of accidents with hydrocarbon products. To reduce the number of accidents in a more evident way, additional measures are necessary.

Legislation about child resistant packaging have to be supplemented with powerful information to the general public, to increase the awareness of the risks associated with these products, and how to handle petroleum products in all forms more safely.

CONCLUSIONS

Poisoning incidents with petroleum distillates are common and seem to increase for every year.

All incidents do not lead to intoxications, but some of them can meet with serious complications and even death.

Child resistant closures can not be considered perfectly safe, and moreover, a large number of incidents occur with products transferred to other containers.

Thus, legislation about packaging must be supplemented with information to increase the awareness of the risks with petroleum products in all forms.

References:

Åkesson J, et al. (1990). Ingestion of aliphatic hydrocarbons can have life-threatening pulmonary effects in small children. Läkartidningen 87:4420-3. (A Swedish publication with English summary and references)

CLINICAL NEUROPSYCHOLOGICAL ASSESSMENT OF PATIENTS CHRONICALLY EXPOSED TO ORGANIC SOLVENTS

Paul G.J. Ganzevles, Bianca W.J. de Geus.

Dept. of Neuropsychology and Psychobiology, University of Limburg, P.O.Box 88, 6200 AB, Maastricht, The Netherlands.

INTRODUCTION

Investigations into the effects of short-term exposure to organic solvents revealed no harmful effects provided that the concentrations were relatively low. However, long-term exposure to low concentrations of organic solvents may affect physical and mental health. After long-term exposure to organic solvents, many workers make strong complaints about headaches, fatigue, forgetfulness, concentration and attention. There are, however, two problems which obstruct the recognition of the effects of long-term exposure: 1) in most cases medical assessment reveals no physical cause for the complaints (Morrow et al, 1990), and 2) until now, epidemiologic research with industrial workers has not produced clear evidence of the toxic effects of chronic exposure to organic solvents (Cherry et al. 1985; Triebig, 1988; Gade et al. 1988). The failure to recognize the complaints, however, causes the patients severe psychological problems. Sometimes they are forced to return to their work, often they are diagnosed as suffering from neurasthenia or dysthymia (DSMIIIR, 300.5, 309.82). Although cognitive dysfunctions can be objectified reliably by neuropsychological assessment, the problem is to distinguish dysfunctions resulting from exposure to chemicals from those resulting from brain traumata or diseases such as Dementia of the Alzheimer Type (Eskelinen et al, 1986; White, 1987). However, if long-term solvent intoxication produces a type of cognitive dysfunction syndrome, the dysfunction profiles of long-term exposed patients should be similar irrespective of the severity of the dysfunctions. We therefore analysed the results of the neuropsychological assessments in order to study similarities in the cognitive dysfunction profiles of exposed persons.

METHODS

Subjects: Twenty-two male patients, referred to our clinic with the preliminairy diagnoses: Organic psycho-syndrome (9), neurasthenia (4), depression(1) and unknown (8). All patients complained of fatique, forgetfulness and attention problems. A questionnaire was used to estimate the nature and the extent of solvent exposure and to reveal whether there were other causes for the cognitive problems. Patients who reported such causes were not included in this study. The autoanamnestic information concerning occupation and organic solvent exposure revealed that 15 patients (3 roof layers, 3 house painters, 3 printers, 4 spray painters, 2 technicians) had been exposed daily for at least 5 years to organic solvents in rather high concentrations (roof layers, spray painters) or for longer periods with low concentrations (printers, house painters). These 15 patients formed Group I (Table 1). In contrast to their verbal reports, there was no clear evidence for chronic solvent exposure in the seven other patients. In four patients (a house painter, spray painter, a technician and a photographer) the exposure had been very irregular with very long intervals; two patients (a chemist and a metal worker) reported that they had worked with organic solvents more than ten years ago and they had no complaints at that time; one patient was a laundryman who sometimes used solvents in chemical cleaning processes. These seven patients formed Group II (see Table 1). T-tests show no significant differences between the two groups on age, education and the reported extent of exposure.

Materials: All patients completed the following tests: Modified Rey drawing task (REY), Raven Progressive Matrices, part A (RAV), WAIS Digit span forwards (DS), 15-word learning test (15 W), Memory Comparison Task (MCT), Tower of London test (TLN), WAIS block design (BLK), Modified Trail Making test (TMT), Stroop test (STR), Symbol-Digit Modalities test (SDMT), Luria/Christensen Neuropsychological Investigation (motor functions, perception of rhythms, visual perception, visuo-construction, speech production, and speech perception).

TABLE 1 MEAN AGE, EDUCATION LEVEL AND EXTENT OF EXPOSURE

	Group I		Group II		t-tests
	mean	sd	mean	sd	p
age	40.3	8.9	44.3	14.2	0.518
education	3.6	0.6	4.4	1.0	0.072
exposure	17.5	10.4	18.6	13.1	0.849

Age and exposure in years; education according to Verhage (1964)

RESULTS

Most of the patients of group I performed three of the Luria /Christensen tasks abnormally or slowly. In the motor function test, 9 of the 15 patients had an abnormal performance while 4 performed correctly but at a slow rate. Abnormal performance was also observed with rhythm perception (9) and with visuo-construction tasks (5 and 9). The patients of group II performed on the Luria/Christensen tasks normally.
The means and standard deviations of measured and calculated data of the psychometric tests and the p-values are given in Table 2. Significant differences existed between the two groups for the Raven test (RAV, max=12), the Digit span (DS), the total score of the five learning trials (15W total, max=75), the delayed recall (15W recall, max=15) and the recognition score (15W recogn.,max=15) of the 15 word-learning test, the time scores of the second, third, and fourth subtask, and the total error score of the Memory Comparison Task (MCT), the score on the Block design test with time limits (BLK time,

max=8) and the time scores on the modified Trail Making subtasks B (TMT B) and C (TMT C).

TABLE 2 MEAN SCORES ON PSYCHOMETRIC TESTS

test	Group I(n=15) mean	sd	Group II(n=7) mean	sd	t-tests p
RAV A	10.4	1.5	11.4	0.8	0.045
DS	4.7	0.8	5.7	0.9	0.045
REY simple	65.5	15.9	52.5	12.6	0.076
REY complex	93.6	46.3	79.3	25.8	0.392
REY diff.	26.3	35.9	23.0	30.1	0.837
15 W total	31.8	6.8	45.0	7.6	0.003
15 W recall	5.4	1.9	9.0	3.1	0.022
15 W recogn.	9.8	4.4	14.5	0.8	0.001
MCT 1	47.0	25.0	37.1	10.2	0.204
MCT 2	65.1	26.5	47.4	4.1	0.024
MCT 3	80.1	32.7	59.7	5.3	0.032
MCT 4	91.8	33.7	71.1	9.4	0.041
MCT errors	11.1	6.9	4.0	2.7	0.002
TLN	8.3	2.4	9.8	2.1	0.191
BLK time	3.5	2.9	7.0	1.1	0.001
BLK number	7.7	0.7	7.8	0.4	0.428
TMT A	55.7	47.7	38.0	7.3	0.182
TMT B	66.8	47.6	30.8	5.4	0.011
TMT C	108.5	100.3	51.1	14.2	0.046
TMT int.	75.9	48.9	47.5	25.6	0.089
STR I	66.7	27.1	51.8	6.4	0.061
STR II	84.1	31.4	66.3	8.3	0.055
STR III	145.8	50.3	114.0	25.9	0.066
STR int.	97.7	43.7	93.9	43.0	0.855
SDMT writing	29.0	10.1	34.5	3.9	0.091
SDMT reading	34.2	12.7	38.6	9.5	0.385

DISCUSSION AND CONCLUSIONS

Patients of Group 1 showed cognitive dysfunctions in a number of tests, in particular memory tests and tests in which speed is important, which is in line with the findings of the qualitative assessment. Differences in test results between groups of exposed and unexposed persons, however, have been reported before (Eskelinen et al, 1985; White, 1987; Morrow et al., 1990) and more data will not help us further unless we look at the data from another viewpoint. A measure for interference can be calculated from the time scores of the TMT and the Stroop test (I= C-B). This interference is viewed as a measure for frontal functions, often called planning or executive functions. As the absolute interference, however, increases with increasing time scores, a better measure for planning functions is the relative interference:

$$I=(C-(A+B)/2)/(A+B)/2)*100.$$

The time scores in the TMT and the Stroop test (Table 2) were different between groups for each subtask, but the relative interference was not different. This effect has also been reported for comparisons of solvent exposed persons with healthy controls (Ganzevles and Houx, 1991). Thus the planning functions of solvent exposed persons may be normal while the speed of information processing is impaired. The results of the other psychometric tests are in accordance with such a hypothesis. In the MCT, Group 1 needed more time and made more errors with increasing items to be processed. The performance of Group 1 was similar to that of Group 2 in the SDMT when the answers were given verbally while Group 1 is inferior to Group 2 when the answers had to be written down, which is an extra activity in this task. Group 1 was inferior to Group 2 on the Block design test when the time limits according to the WAIS-norms were used; without the time limits, the groups performed equally well. The significant differences between the groups on Digit Span can also be explained in terms of slow information processing, which is in agreement with the results of the Luria rhythm perception task. The difference between the groups in the Raven tests was mainly because of the last two items, in which two aspects have to be processed simultaneously.

In conclusion, the differences between the performance of the groups may be the result of a decrease in the quantity of information processing (caused by chronic solvent exposure) while the quality of the performance is in principle spared. Compensation for these disturbances is at least partially possible if the subject makes a conscious effort. The necessity of being continuously alert may be the main cause of the neurasthenic complaints of these patients.

The present results suggest that neuropsychological assessment is valuable for objectifying complaints which are related to long-term exposure to organic solvents; moreover a neuropsychological examination makes it possible to differentiate between exposed and non-exposed persons.

REFERENCES

Cherry N, Hutchins H, Pace T, Waldron HA (1985) Neurobehavioral effects of repeated occupational exposure to toluene and paint solvents. Br J Ind Med. 42: 291-300.

Eskelinen L, Luisto M, Tenkanen L, Mattei O (1986) Neuropsychological methods in the differentiation of organic solvent intoxication from certain neurological conditions. J Clin Exp Neuropsychol 8: 239-256.

Gade A, Mortensen EL, Bruhn P (1988) "Chronic painter's syndrome". A reanalyses of psychological test data in a group of diagnosed cases, based on comparisons with matched controls. Acta Neurol Scand. 77: 293-306.

Ganzevles PGJ, Houx PJ (1991) Psychometric assessment of cognitive dysfunctions after longterm exposure to organic solvents. Eurotox Congress, Maastricht

Morrow LA, Ryan CM, Hodgson MJ, Robin N (1990) Alterations in cognitive and psychological functioning after organic sol vent exposure. J Occup Med. 32: 444-450.

Triebig G (1988) Occupational Neurotoxicology of organic solvents and solvent mixtures. Neurotoxicol Teratol 11: 575-578

Verhage, F. (1964) Intelligentie en Leeftijd. Thesis Assen.

White, R.F. (1987) Differential diagnosis of probable Alzhei mers's disease and solvent encephalopathy in older workers. Clin Neuropsychologist 1: 153-160.

Risk Assessment

(Chair: R. Kroes, Utrecht)

RISK ASSESSMENT - SCIENTIFIC ASPECTS

A.D. Dayan

Department of Toxicology, St. Bartholomew's Hospital Medical College, London EC1 7ED, GB

Most toxicologists now probably accept the paradigm on which their work of toxicity testing is generally considered to be based (Fig. 1).

The accepted definitions (see references listed in Fig. 1) are of hazard as the potential of a physical or chemical agent to do harm, and of risk as the probability that harm will be done under a given sent of circumstances. Accordingly, the experimental toxicologist's work will be heavily weighted towards the study of hazard in toxicity tests and factors that may affect the validity of the extrapolation that underlies the prediction of risk. The clinical or observational toxicologist (or the corresponding environmental toxicologist for eco-systems) concentrates more on investigating revealed toxicity in the clinic (or field) and discovery of the rules of retropolation, so that a better estimate of risk can be made in subsequent studies.

Risk management is the apparent public face of toxicology although in reality it is far removed from the everyday understanding of the man in the street. It requires economic, political and cultural decisions that should involve toxicologists as citizens but not as specific experts. As such, it will be mainly left to others to discuss in this symposium, but it is important to realise that it is usually based on "official" (governmental) acceptance of the same paradigm and its application to much the same data according to similar concepts.

The work of toxicologists is concentrated on recognising and defining risks additional to those of everyday life. It is not always realised how improbable is our existence in the first place and how chancy is our continued being, because of the many natural hazards and diseases that affect all living systems, and the wide variety of serious and even fatal risks that we uncomplainingly (and sometimes unknowingly) accept, and so how irregular is our understanding and governance of these uncertainties (BMA, 1990).

What principles and problems guide and worry experimental and clinical toxicological scientists in dealing with hazard and risk?

Principles underlying Toxicity Testing

In relation to determining hazard and risk, much of what toxicologists do comprises 'toxicity testing', the ordered performance of a sequence of general surveillance studies, which are intended to reveal important hazards. The structures of the tests and

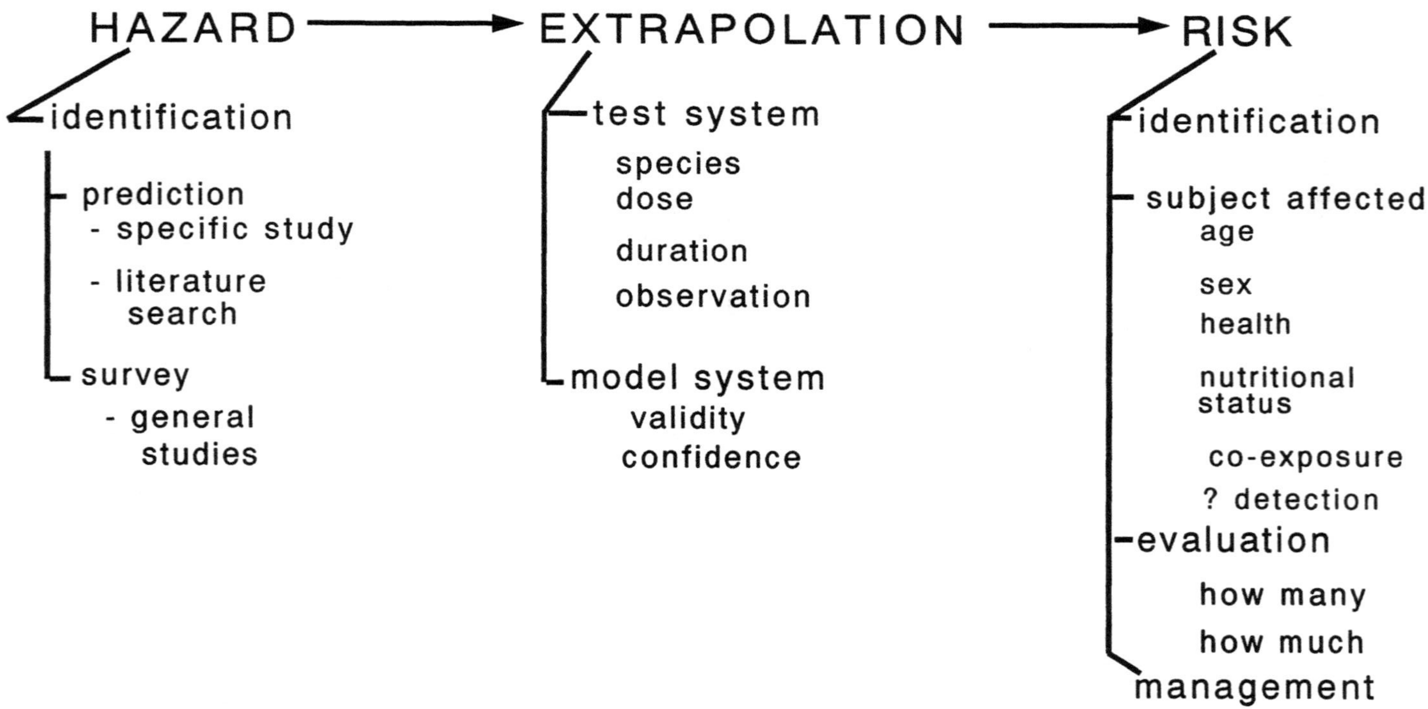

[after Royal Society, 1983
and NAS, 1983
and influenced by BMA, 1990]

principles for interpreting the results have become established by accretion over many years.

Strong control over what practical toxicity testing is done is exerted by reality and belief about regulatory requirements. The guidelines of, for example, the CPMP for pharmaceuticals in the EC, those of the EPA for pesticides in the USA, and of other regulatory agencies, have an enormous influence, and direct almost all the practice of toxicity testing by specifying the nature of the information required for product registration or regulation and how to obtain the data (Alder and Zbinden, 1988). The study protocols and GLP polish of OECD are a further set of constraints, despite carrying pragmatic benefits. Taken together, these 'norms' amount to a monolithic system of investigating substances for many potential types of toxicity - provided that it can be expressed within the expected sorts of experiment done, and that the observer is sufficiently alert to detect both conventional and the possibly more subtle hints of non-conventional effects. Such systems are now so comprehensive that it may seem unlikely that any significant form of toxicity could be missed, but that reassurance still depends on the interpretative power of the scientist (see below). The entire process has the benefits of considerable breadth of coverage and standardisation, but it increasingly risks being driven by more or less mechanical rules to provide material with which it is expected that hazard can be evaluated and decisions made about safety. The sceptical toxicologist will ask all the time whether conventional techniques can encompass actions of types that it has only just become possible to envisage, e.g. immunomodulation, behavioural and other effects on cognitive development. This is not a plea for the application of science fiction or fear to toxicology, but more a pointer to the need for controlled imagination, enquiring whether mechanisms revealed by advances in basic knowledge could be revealed as important toxic processes by current methods.

Commonality of methods is an international goal, but the more difficult interpretation of results often remains at the national level, or is the subject of multilateral debates. It may be a reflection of the power of the human mind, or, pejoratively, an indication of our naïvety, but a striking feature of practical toxicology is the wide agreement about risk and its obverse safety reached by diverse national groups, working at different times, and according to their own national precepts and prejudices.

The observational toxicologist is less constrained by guidelines in detecting and studying manifested risks, as that work is based on the inherent variability of free-living people, animals or natural processes. The system of enquiry depends more on detection of abnormality, the fortunate recognition of a type or pattern of effects that indicates the likely cause or process, and subsequent studies to prove the pathogenesis.

Extrapolation by the experimentalist and retropolation by the field scientist move in contrary but parallel directions, as both attempt to apply general rules to evaluate current problems and to extract new principles to underpin subsequent analyses.

Fundamental features of regulated toxicity testing of this type are reliance on standardised methodology to explore perpetually novel problems, and dependence on analyses by individual experts to interpret findings. These factors together make it less a science than a profession and a craft, in which the personal qualities of the practitioner are necessary to overcome the inadequacies of the data and our limited understanding of the biological processes involved.

The Problems of Toxicity Testing and Hazard Definition

At this level, toxicology comprises set procedures applied to diverse problems followed by mixed analysis of the positive and negative findings according to general knowledge and individual experience.

The alert toxicologist should be able to recognise the occurrence of most hazards, apart from those of an entirely subjective nature, or which are rare, are due to species-specific processes, or are dependent on interactions between systems and populations in the field. None of these can be modelled in the necessarily simplified artifice of the laboratory.

The concerns of the self-critical toxicologist at this level reflect fundamental and methodological aspects of the problems being studied and the methods that have to be employed. Thus, they are partly inherent in the process of the science and arise in part from the accepted working practices.

Methodological questions, as outlined in Fig. 2, are extensively discussed for the experimentalist in monographs on risk assessment and extrapolation (e.g. Clayson, Krewski and Munro, 1985; Roloff, 1987). The observational scientist appears to be less directly resourced, except for the unwisely but commonly overlooked critique of epidemiology by Hill (1965).

In the past few years, particularly in carcinogenicity testing, but also affecting more acute and reversible responses, there have been many instances of apparently misleading findings in toxicity tests, i.e. definite lesions have been produced in experiments, e.g. tumours of the liver and lung in the mouse and of the forestomach in the rat, modulatory effects on the immune system, hypo- and hyper-activity of endocrine organs, and etc. Field experience of man and other animals has led to the realisation that these actions have either not been detected at all, or that the underlying dysfunction has not evolved into the same disease, because it has been too small or too short-lived, or because the relevant biological controls have differed. This has led to what I suggest is an unhappy state in such toxicity work, in which studies are carefully done and then either explanations (mechanistically proven and therefore scientifically acceptable) or excuses (at present lacking a scientific basis and so entirely dependent on experience) are offered as arguments why the apparent hazard can be ignored. Examples are given in Fig. 3, which show how some represent areas of remaining ignorance of the fundamental biological processes of our test systems and others reflect practical limitations of feasible experiments.

Many of the explanations and excuses arise from the variability of biological systems and so are inevitable in laboratory studies based on animals and in vitro procedures. They occur and are just as important in the real world of free-living man, animals and eco-systems exposed to harmful agents because they help to determine individual sensitivity, and they may be the source of effects incorrectly attributed to toxicity.

Risk Assessment
General Problems

Experimental	Studies	Observational
? best or customary test system treatment regime		uncontrollable numbers
? most appropriate observations		status of individual
		1st disorder guides study
? additional investigations		? appropriate investigations
? dose-response relationship		irregular exposure
? species differences & homogeneity		find rare events ? idio-syncratic
? power to detect rare event		confounding factors
		species differences

extrapolation → retropolation

Toxicity Testing

General Principles

Experimental ['toxicity testing']	Observational ['clinical & eco-toxicity']
defined	patient or eco-system
test system numbers dose dose regimen duration	exposure often known or estimated
observations nature & timing general	observations fitted to problem or subjects
non-standard investigations	non-standard investigations
controls negative ? positive	controls 'historical' or experience
simple decision rules	complex decision rules
i.e. survey or controlled experiment	i.e. partly uncontrolled study or survey

A major problem of hazard recognition and of risk prediction that has received less attention than it deserves is that of the causes and significance of individual susceptibility and the consequent range of effects in a large population; see discussion by Brain et al, 1987. In essence, we have come to rely on experience to make judgements about the probable importance of risks, and the application of large and arbitrarily chosen 'safety factors' to achieve the goal of 'safety' i.e. exposure without detectable harm. The use of safety factors has been very effective, but they do not guarantee 'absolute safety', and once risk in very large populations is to be considered, then the practical and statistical limitations to predicting risk must be recognised, e.g. see the cautions advanced in proposing 'acceptable' atmospheric concentrations in WHO (1987), and the general caveats of the ACGIH that preface each annual set of TLVs.

Uncertainties in Risk Assessment

a) Quantification

In many ways the model of hazard identification and its extrapolation to risk, given knowledge of the exposed system or population and the extent of its exposure, has served us reasonably well, because there is no clear indication of its general failure. The dread occurrences of harm, such as thalidomide teratogenicity, male sterility due to dibromochloropropane and dust-induced pneumoconioses and asthma can be attributed to previously unknown and unsought types of toxicity, or to disorders which are so poorly understood that hazard identification is very uncertain, notably allergic reactions and developmental damage.

Our safety has been gained by predominantly qualitative rather than quantitative risk assessment, because of the present weaknesses of attempts at the latter (ECETOC, 1987; HSE, 1989). Instead, large and arbitrary safety factors, based on experimental (e.g. ADIs) or observed (e.g. TLVs and OELs) 'No Effect Levels' have been devised and employed as a convenient means of respecting hazard, whilst enjoying the largely harmless use of agents.

That brief statement does not ignore the considerable effort spent on quantitative risk prediction, most notably in the USA, and the academic and regulatory attempts to devise ever more complex mathematical models to allow for all sources of variation in suggesting harmful and safe levels of exposure. Although 'seductive', so far these techniques have commonly appeared to be misleading and inaccurate, because their simplicity has ignored the true complexities of biological systems. (Clayson, Krewski and Munro, 1985; and the fascinating expose in Columbia Journal of Environmental Law 1989). The newer physiological-pharmacokinetic approach seems more promising, but there are few examples yet to show its validity by comparing experimental prediction with epidemiological findings.

The success of the qualitative approach, which is firmly based on expert assessment has been vital for economic and industrial development, but it represents a deliberate move from conventional science to informed judgement, and so is open to misuse and misunderstanding (Jasanoff, 1990). As a process of analysis and logical synthesis it must be based on available facts and an allowance for the ill-defined problems mentioned above. It offers flexibility at the cost of deliberate imprecision, and if the weaknesses are not recognised, there will be no stimulus to do the research to overcome them; for example, how much effort, particularly in the area of exposure of entire populations, is going into defining and solving the problems listed in Fig. 4 in relation to lifelong population exposure to mixed toxicants via air and water pollution and food contaminants?

b) Assessment of What?

Risk is defined as a probability of some harmful event occurring. But what we experience is both that event and all its mildness or severity. Our perception of risk confounds probability, nature and intensity (Slovic, 1987).

Risk assessment procedures are directed to the probability that an event will occur. We lack good means to measure and express intensity of human distress and suffering. For all exposures except medicines it could be argued that that does not matter, because workers, the public or the ecosystem ought not to experience any harmful effect at any level of intensity. That is an inadequate answer, however, because it implies that the 'safe' level of exposure should be set so low that no effect at all is experienced by any one, and yet we might well be prepared to accept a minor degree of an action in return for, say, employment, e.g. many people accept night and weekend work, despite the cost of antisocial hours, laundry and restaurant workers accept a minor degree of skin irritation of the hands, and those who live at high altitudes accept some limitation of physical performance. In the case of medicines the choice may be more stark - most cytotoxic drugs cause severe nausea and vomiting, which we accept in return for relief and cure of cancer, and hypertensive patients on ACE inhibitors accept distortion of taste because of the benefits of preventive treatments.

Toxicologists and others in most instances have yet to attack the problem of measuring and assessing the quality (nature) of the risks they study. Harm is a vital component in the quality of life, and, as such, it is closely linked with culture and the methods of the social and economic sciences, and ultimately with philosophy and even religious beliefs. We ought to be considering how to approach the problem of measuring and even trying to predict the quality of harm other than through the black and white illusion of legal damages.

Such profound problems cannot long be ignored. We already worry about the quality of our environment and the aesthetic acceptability of industrialised society. There is a toxicological element to these concerns, as well as fundamental aspects of the human spirit, and toxicologists as citizens and as scientists ought to debate the full nature and limitations of their contribution to the risks of living.

I have suggested here first that toxicologists, and those who use their findings, must recognise more clearly the strengths and weaknesses of their methods of detecting hazard and of predicting risk. Second, by presenting hazard and risk solely as probabilities expressed as single figures we may neglect the occurrence of causes of the individual

Figure 4

Examples of Factors Important in Predicting Risks from the Hazards Detected in Toxicity Tests

Toxicological Explanations	Toxicologists' Excuses
species differences in disposition metabolism receptors physiological processes	consequences of over-feeding trauma IN VITRO imbalance of activating and catabolic processes closed system
high dose effect non-linear kinetics	nutritional imbalance circadian rhythm
small experiment group inadequate power	stress activated endemic infection
inadequate detection method	not occur in test or target species
unable model complex interaction	idiosyncracy
different phenotype genotype	allergic response

variation in susceptibility, whch may be very important in determining the harm experienced by a large exposed population. And third, hazard and risk are defined as probabilities, but we may experience both the occurrence of the event (a probability) and its quality (nature and severity).

It may be correct to state, as a master of logic did "Wovon Mann nicht sprechen kann, darüber muss Mann schweigen," but to recognise the limitation of risk assessment is at least to start to improve it.

References

Alder S, Zbinden G. 1988. National and International Drug Safety Guidelines. MTC, Zollikon.

BMA. 1990. The BMA Guide to Living with Risk. Penguin, London.

Brain J D, Beck B D, Warren A J, Shaikh R A. 1988. Variations in Susceptibility to Inhaled Pollutants. Johns Hopkins UP, Baltimore.

Brickman R, Jasanoff S, Ilgen T. 1985. Controlling Chemicals. The politics of Regulation in Europe and the United States. Cornell UP, Ithaca NY.

Clayson D B, Krewski D, Munro I C. 1985. Toxicological Risk Assessment. Vols I and II. CRC Press, Florida.

Columbia Journal of Environmental Law. 1989. Symposium on Risk assessment in Environmental Law. 14 (No. 2), pp289-600.

ECETOC. 1986. Structure - Activity Relationships in Toxicology and Ecotoxicology: An Assessment.

Hill A B. 1965. The Environment and Disease: Association or Causation? Proc. Roy. Soc. Med., 58, 295-300.

HSE. 1989. Quantified Risk Assessment: Its Input to Decision Making. HMSO, London.

Jasanoff S. 1990. The Fifth Branch: Science Advisers as Policymakers. Harvard UP, London.

National Academy of Sciences. 1983. Risk Assessment in the Federal Government: Managing the Process. National Academy of Sciences, Washington DC.

Royal Society. 1983. Risk Assessment. Report of a Study Group. Royal Society, London.

Slovic P. 1987. Perception of Risk, Science, 236, 280-285.

WHO. 1987. WHO Regional Office for Europe Air Quality Guidelines for Europe, pp7-21. WHO Regional Publications European Series No. 23. Denmark.

Legends

Fig. 1 Basic model of predictive toxicology showing the principal factors involved in the evaluation of hazard and risk, and in the empirical prediction of risk.

Fig. 2 General problems of risk assessment experienced by the experimental toxicologist in devising and analysing laboratory experiments, and by the observational toxicologist (clinician, epidemiologist or ecologist).

Fig. 3 Major features of experimental and observational toxicological studies, concentrating on aspects critical in deciding the probable validity of the results in extrapolation to the general human population.

Fig. 4 Common 'Explanations' and 'Excuses' of toxicologists when suggesting that hazards revealed in experiments do not predict risks in man.

INDUSTRIAL ASPECTS OF RISK ASSESSMENT

H.P. Gelbke

BASF, Department of Toxicology, D-6700 Ludwigshafen 1

Risk assessment not only has to rely on the toxicological data base, but the different exposure scenarios must be taken into account to the same extent. Basically, there are three main categories of exposure:

1. exposure during chemical production;

2. exposure of the applicator of downstream user;

3. exposure of the general population.

These may be roughly characterized as follows: at the production site in principle the highest potential exposure might occur, but the most stringent protection methods and enforcement procedures can be installed; by contrast, exposure of the general population mostly is by orders of magnitude lower, but a reliable and enforceable protection by technical procedures may not be possible. Exposure situations for the downstream user and applicator will be somewhere in-between the scenarios described above.

Risk assessments are usually carried out for the general population, taking into account the large number of people and wide heterogeneity of the population. Therefore, very stringent considerations ware warranted. In contrast, at the workplace risk assessment is often governed by spcific side aspects. Some which are typical of the chemical industry have been outlined below.

- Is a complete toxicological profile necessary for chemicals which are only used as industrial intermediates? Should the resources rather be used for technical safety measures so that extensive toxicological investigations can be dropped?

- Can it be taken into account that the general population comprises the very young, sick and elderly people, while at the workplace grown-up persons with a fairly good state of health prevail? If the answer is yes, what ere the appropriate safety factors for classical toxic effects or virtually safe doses for carcinogenic substances?

- If the most sensitice target is specifically either the female or male reproductive function, can such a sex specifity be taken into account for workplace exposure limits?

- At the workplace the inhalation and the dermal routes are of paramount importance, while the toxicological data mostly stem from oral applications. How can the oral dose be transformed to one appropriate for inhalative or dermal exposure?

How can the local impact of aerosols in the respiratory tract be taken into account, especially with regard to clearance impairments and lung burden overload?

- Should different safety factors be used if the inhalation toxicity is governed by either systemic effects or local irritation or if only an unpleasant odor prevails?

- How should effects after dermal exposure be assessed if they only occur at highly irritating concentrations?

RISK ASSESSMENT : GOVERNMENTAL ASPECTS

Gérard PASCAL - Department of Nutrition, Food and Food Safety
National Institute for Agricultural Research
Research Centre of JOUY-EN-JOSAS - F-78352 JOUY-EN-JOSAS (FRANCE)

INTRODUCTION

I shall begin my speech as though I were talking to a meeting of war veterans !

I remember the research I took part in between 1976 and 1979 in France, as a member of a working group whose task was to define guidelines for toxicological assessment of food additives. This work was to be used as a scientific basis for the French Government to draw up a regulatory text which to this day governs authorization of the use of additives in France.

It was with great interest that we discovered the approach proposed in 1979 by our American colleagues from the U.S. Food Safety Council (FSC), introducing the idea of a "Socially Acceptable Risk", and a quantitative risk assessment methodology.

Our discussions were lengthy and lively, and led to a text entitled "Critical Remarks on the Proposed System for Safety Assessment", which was sent to Dr Wodicka, Chairman of the Scientific Committee of the FSC. The conclusion we came to in this text was as follows : "The French Committee acknowledges the theoretical validity of the approach of our American colleagues in their attempts to define levels of risk, but it registers serious reservations on the use by the FSC of the concept of a "virtually safe dose" (VSD) : considerable uncertainties attend assessment of the VSD on the basis of a study of dose-response relationships. It feels that the FSC's approach should be pursued scientifically, but considers that its application would be premature".

In other words, the French Committee refused to implement quantitative risk assessment and remained loyal to the "Acceptable Daily Intake" approach. The Government then followed suit.

Has the situation changed fundamentally today ? This is the question at the centre of the following discussion.

GOVERNMENTAL REQUIREMENTS

If we begin by considering food safety, which is the area I know better than environment or medicine, it can be asserted that governments need to assess :

- safety of use of additives, of processing aids (including enzymes), of flavourings and of packagings ;
- the potential risk related to the use of new technologies or new culinary practices (micro-wave cooking, for example) ;
- the risk related to pollution and contamination of raw foodstuffs.

All these questions are relatively traditional. Other problems are arising which will lead to the adoption of new approaches :

- those posed by the use of new raw materials ;
- those related to the use of genetic engineering in the production of ingredients ;
- those which concern the appearance of ingredients replacing traditional raw materials, and which could lead to profound modifications in food consumption patterns.

The means of assessing "food risk" is sufficiently close to that used in assessing risks related to the use of medicines or those related to the environment for me, to take food as an illustrative example in the remainder of this text.

CURRENTLY AVAILABLE RISK ASSESSMENT METHODS

Before turning to general methods for risk assessment, we should stress the fact that the idea of a "priority ranking system" (valid for additives, colouring agents, flavourings, etc.) has been introduced. This has been developed over the past decade and practically accepted on an international level by expert bodies. When the number of substances to assess is very large (> 1000 for packaging constituents, > 5000 for flavourings), it consists in classifying them according to the level of risk involved in their use. The cost and time required for expert toxicological assessment justify efforts being focused on the top priority problems for Public Health. This method, which takes into account chemical structures of substances, consumption and exposure levels, and quality and nature of toxicological data, was inspired by the work of the U.S. Food and Drug Administration published in 1982 in the form known to all as the "Red Book" of the USFDA (1982). Although this is not a method enabling risks to be assessed, it has proved very useful in enabling official bodies to define their examination priorities. This system is also used in the field of ecotoxicity (Commission of the EC, 1990).

The number of methods generally available for risk assessment has scarcely changed over the past ten years. Each of the main methods has however been discussed and proposals to improve them have been put forward. We shall briefly go through the scientific principles and aspects, which have already been dealt with by another contributor.

1. Determination of Acceptable Daily Intake (ADI) (or of tolerable daily or weekly intakes)

There is no point in re-telling the story of the events involved in the refining of the concept of the Acceptable Daily Intake. Its main promoter, Professor R. Truhaut, recounted this story wonderfully well during a Congress organised by ILSI-Europe in Brussels in 1989 (Truhaut, 1991).

We shall simply say that the Acceptable Dailly Intake is based on determining the maximum dose with no observable adverse effect (NOAEL) on an appropriate animal species. In extrapolating this to man, a safety factor, which must take account of variations in inter-species sensitivity and variations of individual sensitivity within essentially heterogeneous groups of human consumers, is applied to this NOAEL.

The safety factor most commonly used is 100 ; this is subdivided into two factors of 10, reflecting the variations both between and among species.

By this means we arrive at an Acceptable Daily Intake which meets the following definition "Amount of a food additive, expressed on a mg/kg body weight basis, that can be ingested daily over a lifetime without incurring any appreciable health risk, and is based on an evaluation of available toxicological data" (Scientific Committee For Food, 1991).

This ADI (Acceptable Daily Intake) must not however be considered a physical or mathematical constant but a guide serving to calculate admissible limits of the various chemical agents to which consumers can be exposed.

The ADI can in certain cases only be temporary, if certain toxicological aspects are insufficiently clear, or on the contrary unspecified, when using the substance in concentrations just sufficient to reach the desired goal presents no risks for health.

The safety factors can vary between 10 and 1000-5000 depending on the case. The latter figures, which are very high, have been proposed for carcinogens (Poulsen, 1991).

An excellent analysis of the points for and against this method of determining ADI has been published (Johannsen, 1990).

Most governments use the ADI proposed by the international bodies (JECFA and JMPR-FAO/WHO, SCF of the EC) to set content limits for substances, and in particular in the foodstuffs field.

Although some consider the factor 100 to be very arbitrary, many authors, working from experimental data support the idea that a factor of 10 to take account of variations both between and among species is a good approximation in most cases (cf. Johannsen, 1990).

A.G. Renwick (1991) recently proposed that data relating to the kinetics and dynamics of substances should be introduced into assessment of the safety factor in order to provide a more scientific approach of the ADI.

For the moment, governments and international expert committees have taken very little account of toxico-kinetic and toxico-dynamic parameters in their assessments of ADI.

Although the idea of ADI can be criticised from a theoretical point of view, and improved, it can be asserted, as has Pr Truhaut, "that the use of the ADI concept has made a powerful contribution to protecting the health of people the world over, and oiling the wheels of international trade (Truhaut, 1991).

This concept of ADI is accepted for effects other than cancer and genetic mutations by many governments, and has been extended from food additives to pesticides and environmental contaminants in air and water.

2. Quantitative Risk Assessment

The concept of ADI implicitly supposes that the products considered have a concentration threshold below which no toxic effects are registered. This may be applicable to non-genotoxic carcinogens, but it would seem difficult to apply to genotoxic carcinogens. In addition, the existence of the Delaney amendment in the USA, which prohibited the use of carcinogenic constituents as additives (in food and medicines), led the American government into situations which were difficult to manage realistically, since the fine points of the legislation meant that all non-GRAS food constituents had to be considered additives. This Delaney clause is based on the concept of zero risk choice, a notion which will not stand up to factual analysis.

What was necessary, and primarily in the case of carcinogenic substances, was to find a new approach to risk assessment enabling the concept of "virtually safe" introduced by Mantel and Bryan (1961) to be extended. This would then replace the concept of absolute safety, i.e. zero risk.

This alternative approach can be followed through the use of mathematical models, and leads to defining minuscule risks, risks which are so low that they have no practical effect (de minimus risk).

We shall not describe the various mathematical models which, using effects observed after large doses during animal experimentation, can be used to assess the effects of low doses of carcinogens to which man is exposed, and subsequently the risk presented by such exposure.

If we use the classification adopted by Johannsen (1990), these models can be subdivided as follows :

- linear models,
- mechanistic models,
- tolerance distribution models,
- time-to-tumour models,
- biologically motivated models,
- pharmacokinetic models,
- physiologically based pharmacokinetic models.

A very good analysis of the points for and against these models was made by Johannsen (1990).

We shall confine ourselves here to some remarks relating to the implementation of these methods of quantitative risk assessment :

- chemical induction of cancer is a multi-stage process in which one can distinguish initiation, promotion and progression. The first and third stages imply genetic events, and the second epigenetic events. Two classes of genes are concerned : protooncogens and tumour-suppressing genes. The first are activated by qualitative or quantitative alterations due to point mutations, to chromosome rearrangements or to amplification of genes. The latter are, inversely, deactivated through loss of chromosomes, through deletion of chromosomes or genes, through recombination, through gene conversions, or through point mutation (Barrett and Wiseman, 1987).

A remarkable review of the genetic events involved in induction of cancer has just been published by Bishop (1991). The progress of knowledge in this field is spectacular : the complexity of phenomena is brought out to an even greater degree. In particular, it is now apparent that the transformation of normal cells into neoplastic cells requires that several oncogenes act in concert (Hunter, 1991).

Fearon and Vogelstein (1990) published in 1990 a model which can be proposed from current knowledge, to describe the chain of genetic events leading to induction of colorectal tumours (Fig. 1). This is a very good illustration of the multistage nature of cancer induction.

Non-genotoxic carcinogens, i.e. promoters, appear to act by other means. If Ames and Gold (1990) are to be believed, most substances which produce carcinogenic action in animals when administered using the maximum tolerable dose stimulate mitogenesis under these conditions and, thereby, increase mutagenesis. The promoter effect can thus be considered a threshold effect (that below which no more mitogenic effect is recorded). Risk assessment of these promoters can thus be envisaged within the terms of traditional food toxicology (KROES, 1987) :

- choice of the ideal mathematical model supposes a perfect acquaintance with the mechanism of action of the carcinogenic substance. The distinction between a genotoxic substance and a substance with an epigenetic effect is particularly important. But which model should be chosen for a genotoxic ? How many genetic events must occur for cellular transformation to be observable ?

- the double extrapolation from high doses to low doses and from experimental animal species to man also supposes that the mechanism of action are the same in all cases (high and low doses, animal and man). How often do we have all this information ?

These observations show clearly that quantitative risk analysis remains an arbitrary process from many points of view, and thus controversial. All the main American agencies have however adopted this methodology to take decisions relating to the consequences of potentially carcinogenic agents for public health. This methodology is applied to radiation, air pollution, water contaminants, protection of consumers and workers, and management of dangerous waste. The threshold of 1×10^{-6} has been chosen (one lifetime excess cancer risk per one million people) beneath which the impact of carcinogens is considered negligible (de minimus risk).

For the moment, the regulatory bodies in the USA are practically the only ones to use this quantitative assessment method of carcinogenic risk.

Figure 1 : A GENETIC MODEL FOR COLORECTAL TUMEROGENESIS

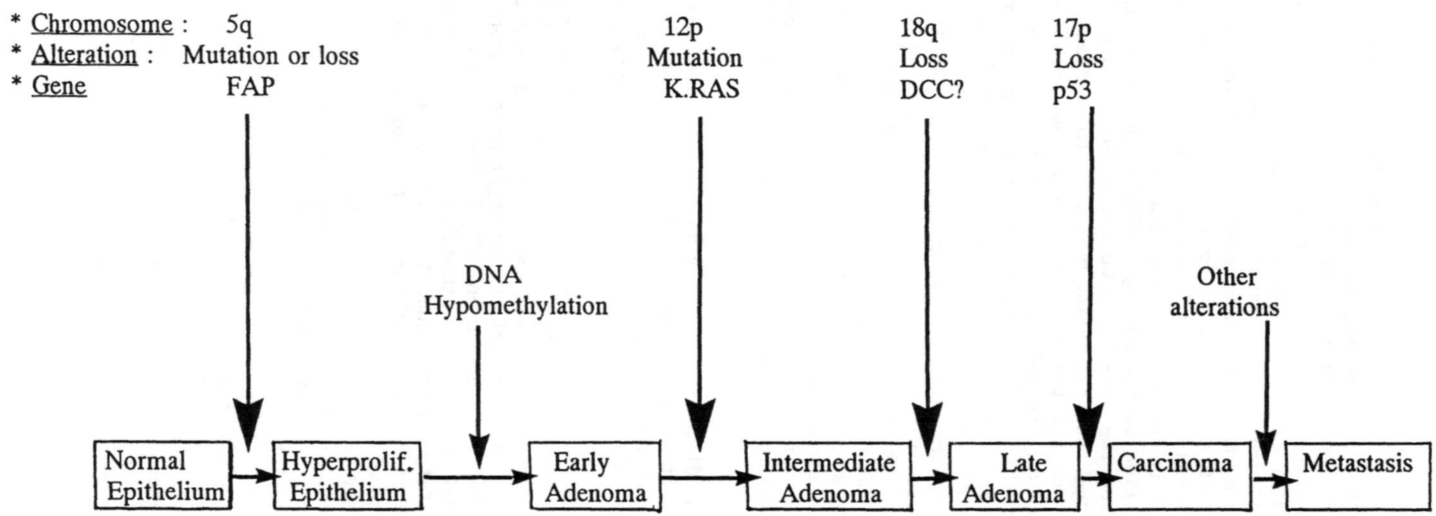

from E.R. Fearon and Bert Vogelstein, Cell, vol.6, 759 – 767, June 1,1990.

CONCLUSION

1. Some aspects of risk assessment have been adopted by international consensus and the corresponding methodologies have been or almost certainly will be very rapidly adopted by governments :

- This is the case with the phase preceding risk assessment. Faced with the size of the task of assessing potentially dangerous constituents, the first step of the official bodies consists in establishing a very approximate classification enabling them to draw up examination priorities. This step is based on the notion of a "priorities ranking system" and is currently the subject of wide international debate, the aim of which is to harmonize the methodology and adapt it to special cases. Such discussions are taking place, for example, in relation to flavourings, within the Codex Alimentarius, the Council of Europe, and the Scientific Committee For Food within the EEC.

- Having defined the examination priority, the process of assessing the risk presented by non-carcinogenic substances has also been agreed by international consensus. The Safety Factor and ADI concepts have been used successfully for a third of a century by the agencies responsible for regulation. In addition, the relative ease with which the safety factor can be implemented, the fact that it is understood by the public, and the caution resulting from its application justify its continued use. It even seems that it could be extended, with some precautions, to risk assessment of non-genotoxic carcinogens. There remain, however, some disparities in quantitative assesment of certain ADIs by the international expert Committees, in particular in regard to food additives (Poulsen, 1991). These are due to differences of opinion as to what should be considered an observable adverse effect, and as to what safety factor should be chosen. These disparities are not very great and should be eliminated as soon as additional toxicological information is made available.

2. The most difficult point remains the assessment of the risk presented by genotoxic carcinogenic constituents. The quantitative risk assessment proposed and applied by the USA remains very controversial. There are many criticisms of this approach, and it is worth recalling some of them :

- the mathematical models chosen rarely take into account the biological events in question. Nor can they take into account qualitative data ;

- the official agencies usually adopt the most conservative models such as the linearized multi-stage model or the Logit model, adapted to a greater or lesser degree. This leads Johannsen (1990), for example, to consider that "the quantitative risk assessment methodology could be viewed as merely an extension of the Safety Factor concept where the size of the uncertainty factor is, in such a case totally arbitrary, and could be as high as five to six orders of magnitude".

Governments may be tempted fo use this approach :

- because they are attracted by approaches which produce the highest level of risk ; this cautious attitude is understandable when subjects relating to public health are concerned ;

- because quantitative risk assessment can seem in the public to be a very scientific (mathematical) approach and hence very serious. The limitations of this argument have been demonstrated ;

- because assessment of very high level of risk can lead some governments, which have reached a very high level of technological quality and of environmental pollution control, to take extremely restrictive regulatory measures. These measures, which would purport to be measures for the protection of public health, could often simply be protectionist measures forming obstacles to the movement of products.

Scientists consider systematic and dogmatic application of quantitative risk assessment dangerous. An interesting illustration of such dangers can be found in the very animated general discussion which came at the end of a conference organised by the Bandbury Centre in 1986 (Bandbury Report, 1987), which was devoted to non-genotoxic mechanisms in the induction of cancer.

The very critical position of the scientists, which even leads some to propose that no decisions should be taken until more knowledge is available, could make us very pessimistic about the chances of finding an international consensus in this field.

I do not share such pessimism fully. Governments are obliged, on an almost daily basis, to take decisions on genotoxic constituents and they cannot wait for scientists to clear up everything under the sun ! I am in agreement with Professor Lave, a professor of economics, who published a wonderful article in Science (Lave, 1987) on risk analysis and, along with him, I feel that "the risk managers need unbiased information with the uncertainties displayed explicitly to help them arrive at good decisions".

Constant dialogue between those who are involved in risk control and specialists in risk analysis is essential to stimulate scientists into making greater contributions. The risk assessment methodology has the great advantage of providing openness in the various aspects of decision-taking. This enables all the interested parties, and more generally the public, to be shown that there is no such thing as a zero risk, and that there is currently no other method than risk assessment to control risk at governmental level. This will also enable light to be thrown on additional research requirements.

Improvement in the knowledge of the various molecular mechanisms involved in tumorous tranformation, and introduction of pharmacokinetic and physiological approaches based on pharmacokinetics, will enable major improvements to be made to the risk assessment process.

Only openness in discussions and cooperation between scientists and those in government responsible for risk control will enable the required international consensus to be reached, and in particular agreement over assessments, a vital precondition for the free movement of products.

REFERENCES

Ames B.N., Gold L.S. (1990). Too many rodent carcinogens : Mitogenesis increases mutagenesis. Science, 249 : 970-971.

Banbury Report (1987). Concluding discussion in Butterworth B.E. and Slaga T.J. (Eds). Bandbury report n° 25, Cold Spring Harbor Laboratory, p. 38mul.

Barrett J.C., Wiseman R.W. (1987). Cellular and molecular mechanisms of multistep carcinogenesis : Relevance to carcinogen risk assessment. Environmental Health Perspectives, 76 : 65-70.

Bishop J.M. (1991). Molecular themes in oncogenesis. Cell. 64 : 235-248.

Commission of the E.C. (1990). Scientific advisory committee to examine the toxicity and ecotoxicity of chemical compounds. Activity Report 1984-1988, DG-V and XI - EUR 12964 : 19-20.

Fearon E.R., Vogelstein B. (1990). A genetic model for colorectal tumori genesis. Cell. 61 : 759-767.

Hunter T. (1991). Cooperation between oncogenes. Cell. 64 : 249-270.

Johannsen F.R. (1990). Risk assessment of carcinogenic and non carcinogenic chemicals. Crit. Rev. Toxicol., 20 (5) : 341-367.

Kroes R. (1987). Contribution of toxicology towards risk assessment of carcinogens. Ach. Toxicol., 60 : 224-228.

Lave L.B. (1987). Health and safety risk analyses : information for better decisions. Science 236 : 291-295.

Mantel N., Bryan W.R. (1961) "Safety" testing of carcinogenic agents. J. Natl. Cancer Inst., 27 : 455-470.

Poulsen E. (1991). Safety evaluation of substances consumed as technical ingredients (food additives). Food Addit. Contam. 8 : 125-134.

Renwick A.G. (1991). Safety factors and establishment of acceptable daily intakes. Food Addit. Contam. 8 : 135-150.

Scientific Committee for Food of the E.C. (1991). Report on a first series of food additives of various technological functions (Twenty fifth series). C.E.C. - EUR 13416 : 4.

Truhaut R. (1991). The concept of acceptable daily intake. Food Addit. Contam. 8 : 151-162.

U.S. Food and Drug Administration (1982). Toxicological principles for the safety assessment of direct food additives and color additives used in food. Bureau of foods, Washington D.C.

U.S. Food Safety Council (1978). Proposed system for food safety assessment final report of the scientific committee. Food Cosmet. Toxicol., 16 (supp 2) : 1-136.

COMMUNICATING ABOUT TOXICOLOGICAL RISKS

R.M. Meertens, G. Kok & T.B. Lenderink

Department of Health Education, University of Limburg, P.O. box 616, 6200MD Maastricht, The Netherlands.

INTRODUCTION

The way experts perceive certain toxicological risks often differs considerably from the way the public perceives these risks. For example, people worry about the health consequences of additives in food, of soil contamination and air pollution, in cases where experts consider the risks very small.

In this paper we will try to explain such differences in perception, and formulate recommendations for the way toxicological risks should be communicated to the public.

Some explanations for the differences in risk perception between experts and laymen are simple, but plausible. First, the opinions of experts may sometimes not even reach the public, and so the public has no other choice than to base their risk judgements on their own or other layman's intuition.

Second, even when the message of the expert reaches the public, the public may choose to rely on their own intuition, because they distrust the expert's message. Lack of confidence in experts may stem from several sources. For example, the experts may have lost their credibility because they were mistaken another time. Sometimes people may even loose their confidence in experts in general, because of a faulty expert analysis in one area. To take another example, disagreement among the experts themselves may also lead people to conclude that experts are not to be believed, that they 'don't know it either'. Indeed, when one is not familiar with the way risk assessments are made, it is difficult to understand why risk judgments differ between experts as much as they do (for example, Californian experts estimate that a lifetime exposure to .007 picograms dioxin per kg a day lead to one additional cancer in 1 million people, while Canadian experts assume that as much as 10 picograms are needed for the same result (Kamrin 1991)).

However, the 'gap' between experts and the public also seems to exist when the public receives the experts' message and considers the experts trustworthy. This paper discusses a third explanation of differences in risk perception between laymen and experts: the way lay people perceive probabilities and risks. The rest of this paper will be mainly devoted to these last topics, as well as how ways people normally assess risks should influence the practice of risk communication.

We will first explain briefly (because this probably will be known by the reader) why experts often cannot be sure of their risk assessments, and why this poses a problem to risk communicators. We will then discuss the way people deal with probabilities, and what they mean when they talk about risk. The last part of the paper contains suggestions on how one should communicate about toxicological risks, given the uncertainty of the experts' risk assessments and the way the public conceptualizes risks.

UNCERTAINTIES IN TOXICOLOGICAL RISK ASSESSMENT

In general, experts base their assessments of toxicological risks on data from experiments with animals. Scientist, however, realize very well that the physiology of man may be quite different from that of experimental animals, and that humans may differ considerably in their sensitivity to a toxin. Therefore scientists use considerable safety margins in their recommendations. The safety factor most commonly used is 100; this is subdivided into two factors of 10, reflecting the variations both between and among species. Furthermore, experts also try to get data on humans in epidemiological research. Of course, often all the evidence of epidemiological and animal studies isn't available, and the only thing experts can do is to make 'educated guesses'.

However, even if all attainable empirical evidence is available, one cannot be sure that a certain substance is causally related to a human illness: while interpretation of experimental studies is difficult because they for ethical reasons are done on animals and not on humans, epidemiologic studies often have an observational, non-experimental nature, and therefore cannot give insight in the causality of relations. Thus, even if all the evidence of animal and epidemiologic studies on a certain potential toxin is in, one cannot give absolute certainty that the substance is (not) toxic.

Some caution in risk communications is even more warranted because of some other shortcomings of risk assessment methods. First synergism or interaction effects may occur (for example, the negative health effects of asbestos are especially strong for smokers). Second, long-term effects (for example next generations), are difficult to determine in research (cf. DES-hormone). Third, though in general adequate, in some cases the safety factors used to take into account inter- and intraspecies variations may not be appropriate (cf. Pascal, 1991).

In conclusion, communications about risks often are to be presented with ample caution by the experts. This (necessary) uncertainty poses a problem for risk communicators: people don't like uncertainty. They call for 'one armed scientists', who can't answer questions with: "On the one hand the evidence is so, but on the other hand..". However, as experts often cannot give the certainty the public wants, they probably better can admit they are not sure: for credibility in the long run it's better to be uncertain than to be falsified.

In the next section, we discuss the layman's perception of probabilities. We will see some experimental illustrations of the preference people have for certainty.

PERCEPTION OF PROBABILITIES

Epidemiologist rely on extensive data sets and advanced statistical computer packages to assess the risk of toxins. Of course, 'the man in the street' does not have such data and tools, and therefore relies upon much simpler processes. For instance by using the so-called 'availability heuristic', i.e. by taking memories which have some relevance to the problem as a starting point.

Let's present an example of how such a heuristic might work. OBagine a man trying to figure out if colorants in food are toxic. A quick memory search results in two examples of people who always eat these incredible pink cakes and puddings. One of them is a child, the other a woman of eighty-five. As the child looks really pale, and the old woman has problems with her stomach, the man is likely to conclude that colorants are toxic.

Apart from the heavy reliance on small biased data sets (salient, vivid or emotionally memories are likely to come up first), the example given shows another characteristic of the layman's way of data handling that considerably differs from the way epidemiologists handle their data: the man only considers one class of data, namely the category of individuals who have a bad health status and who ingested a lot of colorants, and doesn't take into account any data on individuals who did not ingest a lot of colorants, or have good health. Few people would understand that one would need four categories of data to determine the relation in question: (1) healthy people who consume colorants; (2) unhealthy people who consume colorants; (3) healthy people who do not consume colorants; (4) unhealthy people who do not consume colorants (Nisbett & Ross, 1980).

The fast way by which people determine relations between factors may not be considered 'dumb' without further thought. Though man may be considered a 'cognitive miser' when it comes to intuitive epidemiology, heuristic cognitive processing is quite adequate for situations where evidence is strong and there is little time to think. The last situations mentioned probably have had more importance in evolution than sophisticated 'statistic' processes, which only start to have some value after one has managed to survive relatively simple situations where a quick decision is needed.

Of course, the above described layman approach to correlation would not pose any problem to risk communicators when people would easily change their minds when experts inform them of their risk estimates. However, psychological research shows that heuristics still have an influence on risk judgements when objective statistical evidence has been presented (Hendricks, Vlek & Oppewal, 1990; Van de Kar, Meertens, Kok & Knottnerus, submitted). This is intuitively plausible, a lot of people defend their smoking and drinking behaviour by referring to the old man they once knew, who smoked two packets a day and drank a bottle of boose, and lived healthy until ninetysix.

Furthermore, it seems plausible that associations play an important role in determining risks or probabilities. Because some types of chemicals are dangereous, all chemicals (e.g. ascorbic acid or vitamine C) are perceived as toxins. A complicating factor is that, ones an idea is formed about the riskiness of a substance, these ideas are very resistant to change. Either one is selective in the information one seeks, or one interprets the facts in a way that supports the already existing ideas (Nisbett & Ross, 1980). For example, the accident at Three Miles Island was interpreted by nuclear energy proponents as evidence that nuclear energy is safe (there were no big negative consequences), and by nuclear energy adversaries as evidence that they are dangereous (there had been an accident).

Certainty and Uncertainty

Another characteristic of people's concept of probability is that all probabilities are perceived as being a lot alike. A probability of 30% is not as viewed as different from a probability of 50% as one would expect. However, certainty (a probability of 100%) differs psychologically considerably from a probability of 80%, though in both cases the difference is 20%.

An experiment of Kahneman & Tversky (1979) focussed on this issue. One group of subjects was asked to choose between two options, A and B, where option A meant a 20% chance to win 4000, and option B a 25% chance to win 000. From a rational point of view, subjects should choose option A, because the 'expected utility' of option A (1/5 of 000=00) is more than the 'expected uitility' of option B (1/4 of 000 is 50). Indeed, the results show that 65% of the subjects choose for option A. However, another group in Kahneman & Tversky's experiment had to choose also between options A and B; now option A meant an 80% chance to win 000, option B meant a 100% chance to win 000. A subjective utility approach would again predict a preference for option A (8/10 of 000= 200, a greater sum than 000). However, in this case --and that will be intuitively plausible to the most of us-- 80% of the subjects preferred option B: certain wins are preferred to uncertain wins. Empirical evidence further shows this pattern is just the other way around for losses: in general the certainty to lose a certain amount of money is less preferred than the chance to lose a greater amount.

The next study also shows the strong preference people have for certainty (Slovic, Fisschoff & Lichtenstein, 1982a). In an experiment two groups of subjects were asked if they would decide to vaccinate in a certain -- hypothetical-- situation. One group of subjects were told that 20% of the population was expected to be infected with the disease in question, and that the vaccination would give disease protection to half of them. The second group was told that there were two variations of the disease, which both would infect 10% of the population. The vaccination would give protection against one type of the disease, but not against the other. In both cases the chance of infection with the disease is reduced by vaccination from 20 to 10%. However, in the second case subjects choose more often for vaccination than in the first case: in the second case vaccination gives 100% security against one type of the disease. Therefore vaccination seems more useful than in the first (uncertain) case.

Perception of Probabilities and Communicating Toxicological Risks

What are the consequences of lay man's comprehension of probalitities for the practice of toxicological risk communication? First, one may conclude that people may not be expected to deal accurately with probability aspects of toxicological risk messages. Second, people feel that all probabilities have a lot in common, but make a great distinction between chance and certainty. Thus, when experts say that the chance that a certain type of air pollution will cause cancer is really small (for example: .0000016%), people will sometimes perceive this probability as greater than one would expect. By formulating the message a bit differently, facts may sometimes be conveyed better. People may comprehend better that the danger is really small, when the message states that the particular kind of air pollution augments the chance to die of cancer from 24,7% to 24,70000016%.

In the literature more examples can be found of how the way of presentation influences probability perceptions and decision making (Slovic, Fischhoff & Lichtenstein, 1982b; Tversky & Kahneman, 1981). For example, in an experiment with real patients a therapy was choosen more often when the percentage patients that survived (e.g. 90%) was mentioned, than when the percentage of patients that died during treatment was mentioned (McNeill, Pauker & Sox & Tversky, 1982; cf. Siero, Kok & Pruyn, 1984).

Safety Scales

As a solution for the relatively inferior comprehension people have of the concept of probability, several scientist have suggested the introduction of so-called 'safety scales'. The safety scale proposed by Urquhart c.s. (Urquhart & Heilman, 1984) for example shows by a 1 to 8 scale how safe or unsafe a certain activity is: the higher the score, the safer the activity. For example, smoking cigarettes (a packet a day) scores 2.3, flying 5.9. Numbers are assigned to activities by logaritmic transformations of statistical data. When a certain activity for example costs one life a year per hundred people who perform the activity, the score on the safety scale is 2. When the activity costs one life a year per 100.000 people who perform the activity, the score on the safety scale is 5. However, according to Urquhart, it is not necessary for people to know how the score on the scale is computed; he argues that people also have an idea of what an earthquake of five on the Richter scale means without knowing the backgrounds of measurement and computation.

However such a scale in some cases would give the public more insight into the magnitude of the risk, not all risk communication problems seem to be solved. First, one often has to inform people about risks which are relatively unknown, where a detailed estimate in terms of human lifes lost per year is not possible. Second, comparing risks which considerably differ in a qualitative sense does not make anything clearer. For example, the activities parachute jumping and smoking have such a different character that direct comparisons don't make sense: normally one survives the smoking of a cigarette (the health effects only show up after some time), but a parachute jump one either survives, or not. Third, there are often more reasons to call an activity risky than the deaths the activity leads to only. This third point is related to a fourth: people do not define risk as the probability that the activity leads to death. This last point will be addressed in more detail in the next paragraph.

RISK PERCEPTION

Risk, as defined by the man in the street, does not seem to be a uni-dimensional, but a multi-dimensional concept. By using multivariate techniques several studies show two basic dimensions that seem to be related to the concept of risk (Slovic, Fischhoff & Lichtenstein, 1980; Vlek & Stallen, 1981). A first dimension may be called 'threat', a second one 'observability'. A third dimension that sometimes shows up is 'the number of people that may be affected'. The first two dimensions and the characteristics the factors consist of are depicted in Fig. 1. For example, activities are perceived to be less risky when one may voluntary choose to perform the activity, the activity is relatively well-known, and the consequences are controllable and non-global. Furthermore, risks are considered to be more acceptable when there are many advantages attached to the risks.

*not observable
*unknown to those exposed
*effect delayed
*new risk
*risks unknown to science

*controllable	O	*uncontrollable
*not dread	B	*dread
*not global catastrophic	S	*global catastrophic
*consequences not fatal	E	*consequences fatal
*equitable	R	*not equitable
*individual	——THREAT——	*catastrophic
*low risk to future generations	V	*high risk to future generations
	A	
*easily reduced	B	*not easily *reduced
*risk decreasing	I	*risk increasing
*voluntary	L	*involuntary
*doesn't affect me	I	*affects me
	T	
	Y	

*observable
*known to those exposed
*effect immediate
*old risk
*risks known to science

Figure 1: Underlying dimensions of risk perceptions (results of a factor analysis); each factor is made up of a combination of characteristics, these characteristics are also shown in the Figure (source: Slovic et al., 1982).

The qualitative aspects of the risky activity that affect how big the public perceives the risks of the activity, can lead to several recommendations for the practice of risk communication. First, whenever possible, let persons individually choose if they want to take the risk. Of course, practically this is only possible in the case of certain -- individual-- toxicological risks (e.g. additives to food), and very difficult in others (air pollution). Furthermore, one should not forget to mention the advantages the riskful activity brings with it, especially the advantages it has for the general public, for example as a consumer. Also stress what has been done to control the risk and it's possible negative consequences.

Some authors have suggested that 'catastrophic potential' is the most important determinant of the layman's risk perception. The greater and more global the negative consequences that are theoretically possible, the greater the risk is perceived. On first sight this seems a fairly simplistic and crude way of evaluating risks. Several important aspects of the risk, like probability of the negative consequences and the advantages of the risky activities, are not taken into account. From a 'survival of the species' perspective however this definition may be understood better. In a (micro-)evolutionairy sense it makes quite a difference whether an activity costs 1 life per year during a million years, or whether it costs once in a million years the life of a million people in one big catastrophe. In the eyes of people it makes a difference whether in a worst case scenario the whole world population may die, or just an individual. So, the definition of risk in terms of 'catastrophal potential' may have some value after all.

The research referred to above is merely descriptive, and largely a- theoretical. A theoretical basis of the findings may partly be found in stress theory and research (Lazarus, 1966; Paterson & Neufeld, 1989). Stress theory states that stress originates from something (an event, an activity, a substance) that is perceived in a primary appraisal process as a threat ('a risk'). However, only after an inventarisation is made of the available coping mechanisms (secondary appraisal), and these coping mechanisms don't seem adequate to handle the threat, an individual experiences stress. Thus, stress theory seems to underline the importance of the controllability of the threat or risk and it's possible negative consequences.

In health education 'self-efficacy', a concept related to controllability, has proven its value in predicting people's behaviour (Bandura, 1986). 'Self-efficacy' is the judgement the person has of his own efficacy to perform a certain behaviour; in other words, to overcome and control barriers and difficulties. To augment feelings of self-efficacy and controllability, in risk communications one should give detailed and clear information on how the risk and it's negative consequences can be controlled.

Of course, in the case of some risks it is not possible to give detailed information on how the risk can be controlled. A lot of severe threats can not be solved easily. Wortman & Brehm (1975) integrate two social- psychological theories, 'reactance' and 'learned helplesness', to give insight in situations where a threat exists without clear possibilities to control the threat.

In such a situation individuals will feel a loss of control, and first they will show 'reactance', they will try to regain control in several ways. However, when they don't succeed in regaining the control, feelings of helplessness show up: people do not undertake actions to reduce the threat anymore, they become phlegmatic and stay apathetic even when the threat has become controllable again.

When such a situation exists, one might be tempted to conclude that the public is not worried about the threat, because only a small group shows concern in word and deed. The majority of the people do not seem to take the threat seriously. However, this majority might consist mainly of people who take the threat seriously, but do not see any solution and have decided (either consiously or unconsciously) to repress feelings of threat-related concern.

Janssen (1987) illustrated this in a study on the reactions on the air and soil pollution in Eijsden, a village in the South of The Netherlands, at the Dutch-Belgium boarder. A lot of polluting Dutch, German and Belgian industries are surrounding this village. The opposition of the local population however is small; the majority of the population seems not to bother about the pollution. However, a closer look reveals that this is not the case. Janssen used the so-called 'interview fragments technique' (cf. Tomas & Stallen, 1986) to distinguish between really unconcerned people and people who in fact are learned helpless. The fragments used were: *'According to me, there's no proble at all. As far as risks exist, the industries have them under control. The safety measures are so good, that the probability things go wrong is negligible. With respect to the industry, I feel completely safe.' (unconcerned).

*'Some people say we are subject to great danger. To me this seems a bit exaggerated. In fact I'm not sure what to think of the situation. I don't mean there is nothing to worry about, and wouldn't be a proponent for more industry in this area. But you know, you can't think all day about these things; you can't do anything about them.' (helpless)

People had to indicate which of the fragments approached best how they felt themselves. The 'unconcerned' fragment was choosen by 8% of the poulation, the 'learned helpless' fragment by 48%. A third fragment, meant to identify people who were as well concerned as active, was choosen by 24%.

In conclusion, controllability of the risk seems to be a key concept in risk perception. Several domains of knowledge stress the importance of available coping strategies, though the terms used and the content may slightly differ (cf. Rogers & Mewborn, 1976).

A last interesting characteristic of perception individual risk is that in general individuals think that they are less susceptible to negative events like disease, divorce and robbery than the average person (of course, this is logically implausible). One explanation for this so-called 'unrealistic optimism' is that people only believe what they like to believe. Another explanation is that individuals only think of what they themselves undertake to reduce the risk, and don't realize that others also take measures to reduce it (Weinstein, 1984; Svenson, 1981; Weinstein & Lachendro, 1982).

RECOMMENDATIONS

After briefly having discussed the ways experts assess risks, and rather extensively discussed what laymen mean when they talk about probabilities, we will now formulate some recommendations for risk communication.

First, experts should realize (as they often do already) that in most cases the possibility exists that their risk assessments are faulty. What's more, when they only can give 'educated guesses' their judgements of the risks involved may be prone to the same pecularities as those of laymen. Therefore, some caution in advice seems reasonable. Uncertainty about the conclusions of risk assessment is one of the reasons why one should give the public freedom of choice whenever possible. Another reason is that people will find risks more acceptable when they can voluntarily choose between options.

Second, one should stress the advantages the risk brings with it, especially the advantages for the public.

Third, the measures taken to control the risk should be emphasized, especially the measures that reduce the possibility that negative consequences will occur.

Fourth, the way risks and probabilities are presented need careful attention. It is not possible to give general guidelines for how risks should be presented. A comparison between risks may be helpful, but only when the risk one wants to communicate does not differ too much from the risk it is compared with. Furthermore, one should be especially careful in communicating very small and very large probabilities (because of the psychologially great differnces between certainty and probability), and give a frame of reference when possible (e.g. the chance on dying of cancer anyhow). It may also be useful to check the message for emotion arousing concepts, the mentioning of 'death' for example seems to augment the perceived dread considerably.

The suggestions given above are especially important when one wants to reassure the public, when one likes them to view the risks as small as the experts think they are. However, sometimes the public underestimates the risk involved (e.g. bacterial infections in food; risks of using volatile products like paint or glue without appropriate ventilation) and the risk communicator would like people to take risk reducing actions. In this case one should give clear advices on which actions one should take to reduce the risk. Clear advice may heighten the individual's self-efficacy, and therewith the probability that he or she will act like the communicator advises. Furthermore, because of the phenomenon called 'unrealistic optimism', one should emphasize that everybody may be subject to the negative consequences of the risk, and that one should follow up all the advices to reduce the risk.

Apart from these factors, communicators should realize that more is needed for behaviour change than an adequate perception of the risks involved. Generally, factors like self-efficacy, attitude and social norms are regarded as more important (e.g. Janz & Becker, 1984).

Some authors argue that, given experts' uncertainty, one should make risk communication a two-way process whenever possible (National Research Council, 1989; cf. Vlek, 1990). They have come to this conclusion partly on basis of historic cases in risk communication. These cases seem to indicate that risk communications have been most successful in cases where the experts discussed openly with the public from the start, gave them all the information they wanted, and had an open ear for the arguments and fears of the public. A two-way communication process may be especially recommendable with respect to global, non-individual risks: in such a situation, it is the only possibility to give the public at least some freedom of choice.

In all risk communication processes, experts should of course make sure that they are credible, and that their risk communication attempts reach the public.

At last we would like to point out that the recommendations are formulated on basis of the limited evidence available. This evidence not only has accumulated from studies on toxicological risk or communication about toxicological risks, but from a variety of sources. Thus, recommendations have to be regarded with some reserve. Future research should test if they hold for the practice of toxicological risk communication.

REFERENCES

Bandura A. (1986). Social foundations of thought and action: a social cognitive theory. Englewood Cliffs, NJ: Prentice Hall.

Hendrickx L, Vlek C, Oppewal H. Relative importance of scenario information and frequency information in the judgement of risk. Acta Psychologica (in press).

Janssen MEJ. (1987). Ervaringen van inwoners/sters van de gemeente Eysden met de aanwezige lucht- en bodemverontreiniging. Internal report, Rijksuniversiteit Limburg: Maastricht, The Netherlands.

Janz NK, Becker MH. (1984). The health belief model: a decade later. Health Education Quarterly, 11: 1-47.

Kahneman D, Tversky A (1979). Prospect theory: an analysis of decision under risk. Econometrica, 47: 263-91.

Kamrin MA (1991). Risk communication. Paper presented at the XIV World Conference on Health Education, June 1991, Helsinki, Finland.

Lazarus RS (1966). Psychological stress and the coping process. New York: Mc Graw-Hill.

McNeil BJ, Pauker SG, Sox Jr HC, Tversky A. (1982). On the elicitation of preferences for alternative therapies. The New England Journal of Medicine, 306: 1259-62.

Nisbett RE, Ross L. (1980). Human inference: strategies and shortcomings of social judgement. Englewood Cliffs, NJ: Prentice Hall.

Pascal G (1991). Risk assessment: governmental aspects. Paper presented at the 1991 Eurotox Congress, Maastricht, The Netherlands, 1-4 september 1991.

Paterson RJ & Neufeld RWJ (1989). The stress response and parameters of stressful situations. In: RWJ Neufeld, Advances in the investigation of psychological stress. New York: John Wiley & Sons.
Rogers RW, Mewborn CR. (1976). Fear appeals and attitude change: effects of a threat's noxiousness, probability of occurence and the efficacy of coping response. Journal of Personality and Social Psychology, 34: 54- 61.
Siero S, Kok GJ, Pruyn J. (1984). Effects of public education about breast cancer and breast self examination. Social Science and Medicine, 18: 881- 8.
Slovic P. Fischhoff B, Lichtenstein S. (1980). Facts and fears: understanding perceived risk. In R Schwing, WA Alberts Jr, eds. Societal risk assessment: how safe is safe enough? New York: Plenum.
Slovic P. Fischhoff B, Lichtenstein S. (1982a). Facts versus fears: understanding perceived risk. In D Kahneman, P Slovic, A Tversky, eds. Judgements under uncertainty: heuristics and biases. Cambridge: Cambridge University Press.
Slovic P. Fischhoff B, Lichtenstein S. (1982b). Response mode, framing and information processing effects in risk assessment. In R Hogarth, ed. New directions for methodology of social and behavioral science: question framing and response consistency. San Francisco, CA: Jossey Bass.
Svenson O. (1981). Are we all less risky and more skillful than our fellow drivers? Acta Psychologica, 47: 143-8.
Tversky A, Kahneman D. (1981). The framing of decisions and the psychology of choice. Science, 211: 1453-8.
Urquhart J & Heilman K (1984). Risk watch. New York: Facts on file.
VandeKar A, Meertens RM, Kok, G & Knottnerus A (submitted). Determinants of visiting the general practitioner: An observational and an experimental study compared.
Vlek CAJ (1990). Decision making about risk acceptance. Executive summary. Den Haag, The Netherlands: Gezondheidsraad.
Vlek C, Stallen P. (1981). Judging risks and benefits in the small and in the large. Organizational Behavior and Human Performance, 28: 235-71.
Weinstein ND. (1984). Why it won't happen to me: perspectives of risk factors and susceptibility. Health Psychology, 3: 431-57.
Weinstein ND, Lachendro E. (1982). Egocentrism as a source of unrealistic optimism. Personality and Social Psychology Bulletin, 195-200.
Wortman CB, Brehm JW. (1975). Response to uncontrollable outcomes: an integration of reactance theory and the learned helplessness model. In L. Berkowitz, ed. Advances in experimental social psychology, 8: 178-236. New York Academic Press.

Chronic Toxicity

(Chair: E. FOURNIER, Paris)

HEALTH RISK ASSESSMENT OF LONG TERM EXPOSURE TO CHEMICALS : APPLICATION TO CADMIUM AND MANGANESE

LAUWERYS RR, BERNARD A, ROELS H, BUCHET JP, CARDENAS A., GENNART JPh.

Industrial Toxicology and Occupational Medicine Unit, Catholic University of Louvain, 30.54. Clos Chapelle-aux-Champs, 1200 Brussels, Belgium.

INTRODUCTION

Experimental and epidemiologic studies constitute two complementary approaches to assess the risk resulting from long term exposure to chemicals. The main outcome of experimental studies is the characterization of the pattern of toxic effects induced by chemicals, the understanding of the relationship between their metabolic handling and interactions with target molecules and the identification of their critical effects and the most relevant markers to assess their biologically active internal doses. Extrapolation of dose-effect(s) relationships from animal to human is highly questionable, hence the practice to apply arbitrary safety (uncertainty) factors to the doses without observed adverse effects in animals.

For chemicals or groups of chemicals already present in the environment, an epidemiologic approach can be used which offers the advantage to establish the dose-effect/dose-response relationships directly in humans. This approach, however, has been hampered by the difficulty to evaluate the intensity of human exposure to environmental agents and the limited sensitivity of the tools used to monitor health effects.

The study in humans of dose-effect(s) or dose-response relationships can be carried out with the aid of different indicators of exposure or health impairment (Lauwerys 1991). Exposure to pollutants may be characterized qualitatively (questionnaire, job exposure matrices ...) or quantitatively by environmental (e.g. market basket surveys) or personal (e.g. duplicate meal surveys, personal sampling of airborne contaminants) monitoring or by measurements of the internal dose or the target dose. The adverse health effects may be expressed in terms of increased death rate, clinical diseases, irreversible or reversible functional changes or critical biological changes i.e. changes which are predictive of health impairments if they persist or are recurrent.

Numerous epidemiologic studies on the health effects of environmental and industrial pollutants have focused on the relationships between various external indicators of pollution (presence of emission sources, ambient monitoring ...) and morbidity (clinical signs) and/or mortality data. This type of studies suffers from many limitations. Current and/or past exposure to pollutants is usually estimated on a group basis. Furthermore diagnosed clinical entities are often late manifestations of pathophysiological processes, the evolution of which may have been influenced by numerous exogenous and endogenous factors. With such an approach, environmental and/or industrial pollutants can only be identified as etiologic agents of pathologic conditions when their role is predominant in comparison with that of other factors (e.g. lung cancer in asbestos exposed workers) or when the time interval between exposure and diseases (latency period) is not too long (e.g. Spanish oil syndrome; lung fibrosis in diamond polishers). But in general, exposure to environmental or industrial chemicals is rarely of such intensity that their clinical effects become rapidly apparent. Furthermore, since many diseases are multifactorial, the contributory role of environmental or industrial factors may be easily overlooked with an epidemiologic approach based on a crude assessment of exposure and a search for late effects (diseases, causes of death).

These difficulties can be partly overcome if the assessment of exposure is carried out on an individual basis (e.g. by personal environmental monitoring techniques or better by biological monitoring methods) and if early biological effects are selected as health indicators. In summary, health risk assessment is best performed by epidemiologic studies in which the individual dose (and ideally the target dose) and the critical biological changes are monitored with sensitive and specific markers. But the use of such markers of exposure and effects requires the knowledge of the fate of the chemical(s) in the organism and its mechanism of action (or its critical target organ(s)). As indicated above, these are the fields in which experimental studies are most relevant. So far, an epidemiologic approach relying on the measurement of the target dose and/or the critical biological changes has only been used for a limited number of pollutants. The assessment of the critical exposure level to cadmium illustrates this approach.

CADMIUM

Biological indicators are available to estimate current and life-time integrated exposure (cadmium in blood and in urine) (Lauwerys et al 1976) and to detect the early changes in the main target organ, the kidney (Bernard and Lauwerys, 1990).

In the framework of a collaborative EEC project, we have recently examined a cohort of workers exposed to cadmium and a matched control group for the presence of early renal changes. A battery of biological markers was used in order to detect changes at different renal sites (glomerulus, proximal tubule, loop of Henle, distal tubule, interstitium) (Lauwerys and Bernard 1989). The kidney markers could be classified into three categories 1) functional markers : urinary proteins derived from the plasma (retinol binding protein, β_2-microglobulin, protein 1, albumin, transferrin, IgG) and serum β_2-microglobulin and creatinine, 2) cytotoxicity markers : tubular antigens (BB50, BBA, HF5, intestinal alkaline phosphatase-IAP-, Thamm-Horsfall glycoprotein-THG-) and enzyme (N-acetyl-β-glucosaminidase-NAG-) in urine, 3) biochemical markers : prostaglandins (6-keto-PGF1α , PGE_2, PGF_2) thromboxane, kallikrein, sialic acid and glycosaminoglycans in urine, sialic acid in plasma and red blood cell membrane negative charges.

A multivariate correlation analysis revealed a statistically significant association between the cadmium concentration in urine and the urinary output of albumin, transferrin, β_2-microglobulin, THG, NAG, GAG, tubular antigens (BB50, BBA, HF5, IAP), 6-keto-$PGF_{1\alpha}$, PGE_2 and sialic acid and the concentration of β_2-microglobulin in serum. Most nephrotoxicity markers studied were significantly increased when the cadmium concentration in urine exceeded 10 µg/g creatinine, which supports our previous estimate of the critical concentration of cadmium in urine for the risk of nephropathy in occupationally exposed subjects (Bernard et al 1979; Buchet et al 1980; Lauwerys et al 1974; Roels et al 1981). However, the application of a logistic regression model revealed that the urinary excretion of albumin, NAG, GAG, BBA and THG and the serum level of β_2-microglobulin already tended to increase from thresholds of urinary cadmium between 3 (for albumin) and 5 (for THG) µg per g of creatinine. This indicates that the threshold of urinary cadmium associated with an increased risk of subclinical renal effects may be inferior to the level so far considered as critical.

Furthermore, in view of the well known influence of the "healthy worker effect", this threshold might still be lower for the general population mainly exposed to cadmium via food and tobacco smoking. Indeed, a similar study performed on 1699 subjects non occupationally exposed to cadmium revealed that the probability of tubular dysfunction (as assessed by sensitive tests) was about 10% when cadmium in urine reached 2-3 µg/day, probably corresponding to a renal cortex concentration of about 50 ppm. It can be estimated that in non-smokers this level is attained after 50 years of an oral daily intake of about 1 µg/kg body weight (Buchet et al 1990; Lauwerys et al 1990).

MANGANESE

We have carried out epidemiologic studies on workers to determine the no-adverse effect level of manganese containing dust (Roels et al 1987; Roels et al to be published). For this metal, the situation is less favourable than for cadmium since no biochemical markers have yet been identified to assess the target dose and the toxic effects after long term exposure.

The critical target organ after chronic exposure to manganese is the central nervous system (neurobehavioural symptoms and neurologic signs characteristic of an extrapyramidal syndrome which has several similarities to Parkinson's disease). Since very efficient homeostatic mechanisms prevent large fluctuations of manganese concentration in whole blood and since manganese is mainly excreted by the biliary route, it has not yet been possible to identify a biological marker to assess the intensity of exposure or the concentration in the target organ. In industry, evaluation of individual exposure to manganese is thus best carried out by monitoring its concentration in total and respirable dust in the breathing zone of the workers. Furthermore, no biochemical indicator is available for the detection of the early neurotoxic effects of manganese. Presently, neurofunctional examination (e.g. the measurement of visual reaction time, eye-hand coordination, hand steadiness) represents the most sensitive approach for this purpose.

These indicators of exposure and toxic effects were monitored in two groups of workers exposed to manganese containing dust and their matched controls. Study of the dose-response relationships indicated that the prevalence of some neurofunctional disturbances was related to the integrated exposure to airborne manganese (total and respirable dust). A logistic regression analysis of the data also showed that tremor (hand steadiness) was the most appropriate parameter to define a threshold effect level. This analysis suggested that a lifetime integrated exposure to manganese dust (as MnO_2) above 3575 µg Mn/m³.year (total dust) or 730 µg Mn/m³.year (respirable dust) may lead to an increased risk of tremor. It can therefore be concluded that for a professional life of 45 years, the current occupational exposure limit for manganese (total dust : 5 mg/m³) is too high and should be reduced by about 60 fold in order to protect the majority of workers from the neurotoxicity of manganese.

Unfortunately, for the reasons given above, the risk of overexposure to manganese cannot yet be defined at the individual level through biological monitoring methods and the health surveillance of the workers must still rely on the search for functional changes which, when present, probably reflect an important accumulation of the metal in the central nervous system.

CONCLUSION

The identification of markers of early critical effects and biologically active doses constitutes an important outcome of the studies on the metabolic handling and the mechanism of action of chemicals and is essential for health risk assessment. Epidemiologic studies which rely on these markers for the assessment of dose-effect relationships are the most relevant to suggest acceptable exposure levels.

REFERENCES

Bernard A, Lauwerys R (1990) Early markers of cadmium nephrotoxicity : biological significance and predictive value. Toxicol. Environ. Chem. 27:65-72.

Bernard A, Buchet JP, Roels H, Masson P, Lauwerys R (1979) Renal excretion of proteins and enzymes in workers exposed to cadmium. Eur J Clin Invest 9:11-22.

Buchet JP, Roels H, Bernard A, Lauwerys R (1980) Assessment of renal function of workers exposed to inorganic lead, cadmium or mercury vapour. J Occup Med 22:741-750.

Buchet JP, Lauwerys R, Roels H, Bernard A, Bruaux P, Claeys F, Ducoffre G, De Plaen P, Staessen J, Amery A, Lijnen P, Thijs L, Rondia D, Sartor F, Saint-Remy A, Nick L (1990) Renal effects of cadmium body burden of the general population. Lancet 336:699-702.

Lauwerys R (1991) Occupational Toxicology. In : Klaasen CD, Amdur MA, Doul J (ed.) Casarett and Doull's toxicology : The basic science of poisons, 4th edn. Macmillan, New York, Chapter 29.

Lauwerys R, Bernard A (1989) Preclinical detection of nephrotoxicity : description of the tests and appraisal of their health significance. Tox Letters 46:13-29.

Lauwerys R, Buchet JP, Roels H, Brouwers J, Stanescu D (1974) Epidemiologic survey of workers exposed to cadmium : effect on lung, kidney and several biological indices. Arch Environ Health 28: 145-148.

Lauwerys R, Buchet JP, Roels H (1976) The relationship between cadmium exposure on body burden and the concentration of cadmium in blood and urine in man. Int Arch Occup Environ Health 36:275-285.

Lauwerys R, Amery A, Bernard A, Bruaux P, Buchet JP, Claeys F, De Plaen P, Ducoffre G, Fagard R, Lijnen P, Nick L, Roels H, Rondia D, Saint-Remy A, Sartor F, Staessen J (1990) Health effects of environmental exposure to cadmium. Objectives, design and organization of the Cadmibel study : a cross-sectional morbidity study carried out in Belgium from 1985 to 1989. Environ Health Perspect 87:283-289.

Roels HA, Lauwerys RR, Buchet JP, Bernard A, Chettle DR, Harvey TC, Al-Haddad IK (1981) In vivo measurement of liver and kidney cadmium in workers exposed to this metal : its significance with respect to cadmium in blood and urine. Environ Res 26:217-240.

Roels H, Lauwerys R, Buchet JP, Genet P, Sarhan MJ, Hanotiau I, de Fays M, Bernard A, Stanescu D (1987) Epidemiological survey among workers exposed to manganese : effects on lung, central nervous system and some biological indices. Am J Ind Med 11:307-327.

Roels HA, Ghyselen P, Buchet JP, Ceulemans E, Lauwerys R (to be published) Assessment of the permissible exposure level to manganese in workers exposed to manganese dioxide dust. Br J Ind Med.

ANIMAL STUDIES FOR PREDICTION OF CHRONIC TOXICITY

G H Pigott

ICI Plc, Central Toxicology Laboratory, Alderley Park, Macclesfield, Cheshire, UK

INTRODUCTION

The aim of a chronic toxicity study is to administer the test substance for a long enough period for chronic effects to be realised. It may be equally important to determine the progression of effects seen in shorter term toxicity studies. The number of animals and dose levels should be sufficient to establish dose response characteristics and a no effect level. The range of examinations within the study should cover several parameters of toxicity and, where possible, examinations should be undertaken at intervals throughout the study to follow the development of any chronic effects.

STUDY DESIGN

There is international agreement on the basic design for chronic toxicity studies, although national guidelines differ in detail. The OECD guidelines (OECD 1987) are widely accepted and provide a uniform basis for study design. Most investigations are now performed under the Good Laboratory Practices regulations which ensure some uniformity of data collection/recording.

A recent tendency for some regulatory authorities to regard their guidelines as regulations should be resisted. Guideline studies may be inappropriate in specific cases. Nevertheless, deviations from guidelines should be justified on a case by case basis, even when experiments are not directed to specific regulatory targets. This will help avoid the generation of spurious results with a consequent need for repeat studies to evaluate them.

While guidelines ensure a level of comparability between experimentors, there remain factors of study design and conduct which can lead to apparent inconsistencies. Some of the problems and pitfalls associated with the design, conduct and interpretation of chronic toxicity studies in animals are illustrated below.

SOURCES OF VARIATION

The three basic factors which may influence the results of a chronic toxicity study are, the test substance, the test system and the detection system.

Test Substance:

The intrinsic determination of response. Technical grades of test substance are used in a suitable formulation to ensure stability and to minimise inter-group differences in such variables as dose volume. It is well recognised that formulation may influence absorption rate and this may result in differences in toxicity. The influence of particle size of solids is well documented for inhalation exposures but is less well recognised for other routes. Particle size differences may result in substantial changes in absorption rate from a given formulation and hence influence study results, though usually only quantitatively. When presentation is in the feed the particle size must not be so large as to allow selective feeding.

Test System:

Induced variability may be grouped into three broad areas:species, exposure regimen and environment.

Species: All test guidelines advocate the rat. Other rodent species may be justified on criteria specific to the test substance. Where a second species is required the dog, usually the beagle, is preferred. Primates are also acceptable as a non-rodent species but their use is declining as they offer few advantages of sensitivity/specificity and animal welfare considerations increasingly restrict their use.

Selection of strain is usually determined by custom and practice within the testing laboratory. An outbred strain is often claimed as representative of a larger population and may give an indication of the range of response to be expected, though inbred strains are also widely accepted. A substantial background database is considered frequently to outweigh specific disadvantages associated with a particular strain.

Exposure regimen: An important consideration is the selection of an exposure route. Those available are inhalation, oral or dermal administration with implantation and injection used for some medical/pharmaceutical products. The oral route is frequently advocated for primary testing with others only as a secondary evaluation. This may be a mistake. Both asbestos and crystalline silica are highly toxic by inhalation (and have induced much human disease by this route) but are innocuous materials when ingested. Few dermal tests are performed, though this may be the practical route of exposure for many test substances. Oral administration ensures that the majority of absorbed material passes through the liver before other organs are exposed. Alternative routes may change the distribution of test substance and therefore the balance of toxic effects.

The rate of test substance administration has a clear influence. Sometimes, heroic doses have been used to induce signs of toxicity. These signs may have little or no relevance to the supposed test substance. Thus, sodium saccharin is apparently more toxic than calcium saccharin, though at the doses involved the comparison is between the toxicity of sodium ions and calcium ions. Excess of acidic or basic compounds may induce bladder hyperplasia, irrespective of their nature (de Groot et al, 1988). Bladder hyperplasia possibly leading to tumours, may also result from irritation by crystals of soluble test substance or metabolite after heroic doses, rather than an intrinsic property of the test substance. Fortunately, this is now well recognised and most regulatory authorities accept a maximum dose rate of lg/kg/day as a "limit" test. Where it is genuinely necessary to go to higher doses, e.g. evaluation of a foodstuff, then specific diets formulated around the test substance may be needed to avoid nutritional imbalance between groups.

Where signs of toxicity are evident below the limit dose, most guidelines specify at least three dose levels. The highest of these (MTD) "should show signs of frank toxicity". There is agreement that the MTD should not be associated with "excessive" toxicity which may compromise survival, induce potential nutritional imbalance or produce severe pharmacological or physiological effects which might compromise the study results. There is still debate between those who characterise the MTD as "maximum tolerated" dose, (Farber 1987, Haseman, 1985) and those advocating the "minimally toxic" dose exemplified in the OECD Guidelines. The difference between the two may be substantial. Much of the debate has focused on the carcinogen bioassay, though considerations for chronic toxicity are similar. The use of too high a dose may be misleading in a chronic study. A subchronic study, usually conducted at higher doses, will have established a primary effect. To reproduce this in the chronic study risks masking other more subtle changes which may occur only over a prolonged period. Provided there is an adequate margin of safety over the projected exposures in use, the requirement to produce frank toxicity is arguably excessive.

Selection of the other dose levels is less controversial. The mid-dose may show minimal toxicity or pharmacological change associated with the test substance. Frequently, this is used to define a no observed adverse effect level (NOAEL) where features such as pharmacological effects or (reversible) adaptive hypertrophy are evident. The lower dose is used to define an anticipated no effect level for both toxicity and pharmacological/adaptive change and, as such, is frequently conservative. This latter is important when considering extrapolation of results in risk assessments.

Where test substance is administered in the feed, the effective dose rate will vary with age. **Fig 1** shows a typical curve for the rat. The variation may be a reasonable model for a test substance presented as a constant proportion of the diet, e.g. a pesticide residue on food where the enhanced dose rate seen with younger animals may extend to human exposure. Nevertheless, the effect is exaggerated in experimental animals, particularly during the rapid phase of growth. This should be borne in mind when apparently transient effects are observed or when responses diminish with time. An alternative is to adjust incorporation rates in the diet to give a constant dose rate. This is a feasible design though it is difficult to achieve high precision. Where there is some relationship to absolute concentration in the diet, e.g. a stomach irritant, increasing inclusion rates to counter decreased food consumption may enhance the response such that the effective achieved dose is actually reduced.

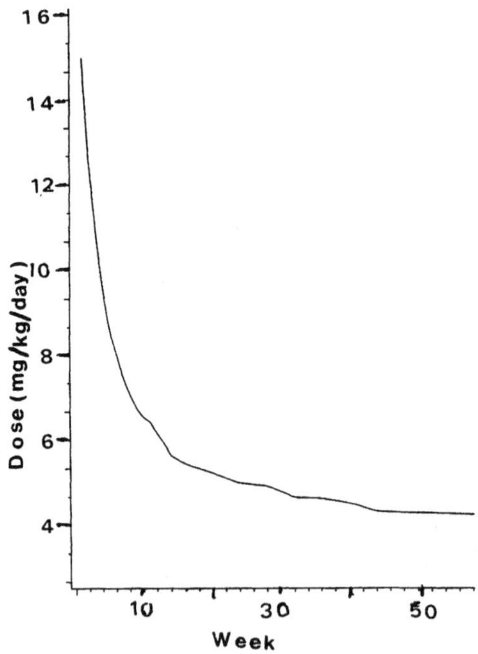

Figure 1 Variation of dose-rate (mg/kg/day) with age for male rats given test substance at 100ppm in diet

A similar effect may be seen with gavage dosing, where significantly toxic doses often induce inappetence. A subsequent dose on an empty stomach may enhance the effect. This exaggerates the diurnal variation in test substance concentration in the blood associated with bolus dosing and can account for wide intra-group differences in response by this regimen which may be wrongly interpreted as reflecting genetic variability in the test population.

Current regulatory guidelines for the duration of rodent studies, specify periods varying between 12 and 24 months. However, several investigators have demonstrated that most signs of chronic toxicity are evident by 90 days and Walker and his co-workers have demonstrated that for a range of pharmaceutical compounds, a six month period is adequate for studies of this type (Lumley and Walker, 1986). For inhalation exposures, especially with particulates, it is not uncommon for change to be progressive and the full extent may not be evident until one year or later. Nevertheless, continued exposure may not be necessary. There is a benefit in reducing the duration of chronic studies, particularly in rodents. Towards the end of the lifespan, geriatric change results in increased variability in both clinical parameters and histopathology in the rat with a consequent loss of sensitivity of these assessments. In addition, group sizes must be increased to ensure sufficient survivors for evaluation. A duration of "at least six months" is also sufficient to demonstrate chronic toxicity in the dog (GIFAP, 1982). Thus, a study duration of one year should prove adequate for assessment of chronic toxicity in all test species, while avoiding the problems of interpretation against a background of geriatric change. It may be claimed that such a regimen would not allow full evaluation of effects on the ageing organism. While geriatric toxicity may require additional investigation, it is not self-evident that this should form part of a chronic evaluation; habituation throughout the lifetime could change the response to an acute dose in old age.

<u>Environment</u>: There is a current fashion to specify close limits for temperature and relative humdity, particularly for rodents although there is little evidence to support the need for this. Light cycle maybe more important. Uneven day length should be avoided and the exclusion of natural daylight is desirable to reinforce the specified cycle.

Housing density is an important factor. Single housing leads to a shortening of the life span in rodents, possibly related to stress and may lead to behavioural disorders in non-rodents. Where single housing is necessary from scientific considerations visual contact with other test animals should be maintained. Gang housing, particularly for rodents, may offer benefits in animal welfare. Providing cage mates are introduced to each other as early as practicable after weaning, there is little problem with fighting.

Dietary factors are known to influence studies. Dietary composition may influence the type and content of cytochrome P450 in the rat liver and thus the innate ability to metabolise xenobiotics. The fibre content of the diet may affect mineral absorption as well as gut motility (and thus duration of exposure).

A more general problem is over-nutrition. It has been demonstrated (Hart and Chou, 1991) that severe reduction of dietary intake (<70% of *ad libitum* value) is associated with increased longevity, decreased spontaneous disease and a decreased response to toxins in rodents. Differences are generally quantitative, though the influence on high dose results should not be overlooked when evaluating a dose-relationship.

Environmental factors generally contribute to a study through their effect on animal health. A disease-free stock and a husbandry regime adequate to maintain this status is most important for the successful conduct of a sensitive chronic toxicity study. Interpretation of subtle change in the presence of background intercurrent disease may not be impossible but is certainly difficult.

Detection System:

The range of measurements made is the most important factor. Where specific end points are of particular interest, it is advantageous to include as many measurements of change as possible. This isolates possibly spurious differences in a single parameter and by reinforcing conclusions multiple end points can improve sensitivity. This may be an important factor. Even with 20/sex/group the ability to detect small changes in clinical parameters and organ weights may be low. This is shown for selected organ weights in the rat in Table 1.

Table 1

	Organ	
Probability of Detecting as 'Statistically Significant" ($p<0.05$) a 10% change in specified organ weights in rats (20/sex/group)	Kidney	66%
	Brain	99%
	Adrenal	33%
	Heart	93%
	Liver	66%
	Lung	66%
	Testes	97%
	Ovaries	24%

This exemplifies the weakness of 'statistical significance' as the sole measure of effect (or lack of it). Nevertheless, statistical considerations are important in overall study design. A randomised block structure in which environmental differences are minimised is a sensible feature of any study. Although not achievable for inhalation studies, the housing of groups in blocks and/or separate rooms should be avoided whenever possible. Where an outbred strain is used, a randomisation process where litters are equally represented in each test group and control, also assists in reducing inter-group variation. This is sometimes at variance with regulatory requirements for a narrow bodyweight range at the start of the study. However, the benefits of reduced inter-group variance in other parameters may outweigh any minor differences in initial bodyweight, as the effects of this can be minimised by appropriate statistical evaluation.

INTERPRETATION

A major aid to sensitivity is the application of optimum statistical methodology. The interpretation of bodyweight differences is a good example. The most important influence on bodyweight in growing rats, is the previous weeks' bodyweight; an efficient statistical evaluation will take this into account. Such analysis, when combined with a visual index of variability, can be used to test significance of particular changes against a pre-set difference determined to be biologically significant. Examples of this presentation are shown; for a (real) inhalation study where the many "statistically significant" differences were in fact spurious (Fig. 2a); for bodyweight divergence due to a small but continuing toxic effect (Fig. 2b) and a palatability/adaptive effect of compound, where there is a transient difference followed by a growth curve parallel to that of control (Fig. 2c). Such analyses derive maximum information from the minimum number of animals and should be encouraged as the pressure to reduce animal usage increases. Unfortunately, there remains a tendency for evaluators in regulatory agencies to apply a single criterion for "significance" (usually $p < 0.05$ by Students t-test) and require explanations for any differences they identify. This can be a time consuming and not very rewarding exercise.

Some parameters, particularly in clinical pathology, may show quite large variations and interpretation of minor change may be diffcult. The range of values historically encountered in the same laboratory for studies conducted at about the same time, may be useful comparators to place isolated minor changes into context. However, the range should relate to individual values and not group means. It is not always valid to argue that because the group mean lies within the background data range, effects are not of toxicological significance. Exclusion of extreme values is also contentious. If they can be associated with specific clinical disease or other identified change, then omission of extremes can help determine whether there is an underlying trend. These values may also be omitted where their magnitude indicates an obvious experimental error, but in other cases they are probably best retained.

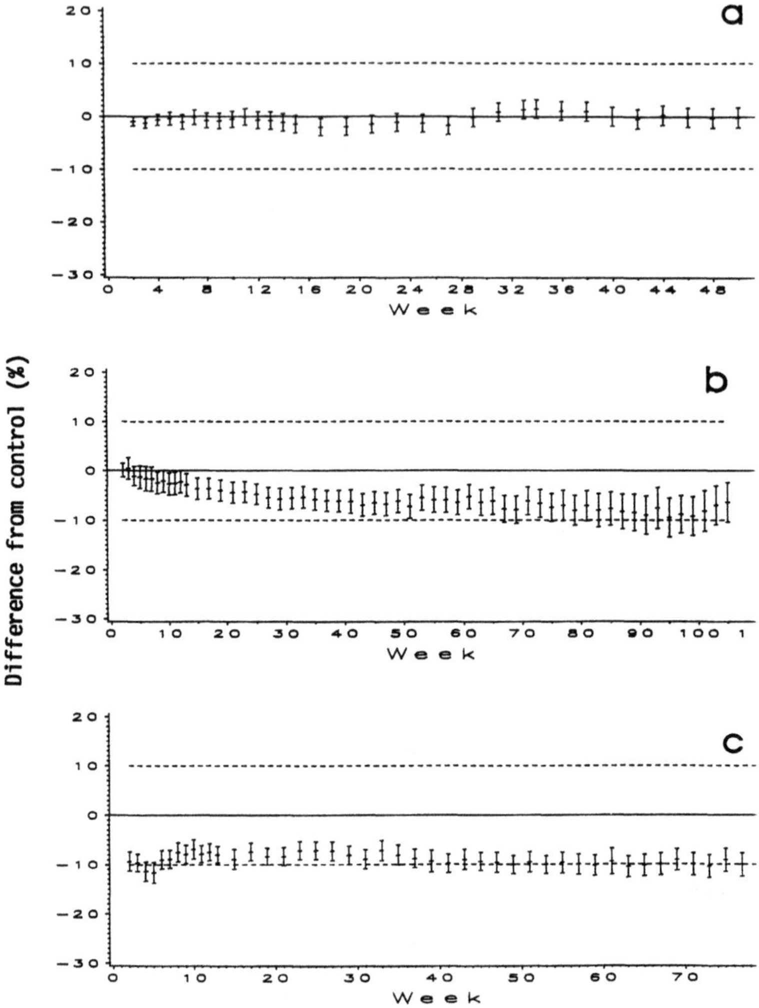

Figure 2 Statistical Analysis of bodyweight adjusted for initial weights
a) for a rat inhalation study at a limit dose (no effect).
b) male mice showing a customary toxic effect.
c) palatability/adaptation in male mice. The dotted lines are at ± 10% of mean, error bars show standard deviation.

Continued debate about significance of minor changes underlines the lack of general agreement on the magnitude of biologically significant change for many parameters. Where views are expressed on this issue they are often at variance with the sensitivity expected from the study design.

RELEVANCE TO MAN

The primary use of chronic toxicity studies is in hazard assessment. A well conducted study can define with some confidence the nature of an effect, demonstrate a no observed effect level (though this is frequently conservative) and give dose response information. Before these data are used in a risk assessment, it is necessary to establish the validity of the study against the criteria mentioned above and then assess the relevance of such observations to man. In many instances, these are well known or readily extrapolated, due to recognised similarities in underlying biochemical mechanisms. However, inter-species differences in metabolism are not rare and unusual observations may require supplementary investigation to put them into context. The first of these is often to repeat the test in a second species. This is probably the most useful application of the second species requirement. If the results coincide, the probability that man will react in the same way is substantially increased. If not, then more detailed investigation of mechanism may be required to establish the significance for human exposure.

Examples of chronic effects which are not relevant to man are accumulating. Some are quite specific, such as nephropathy due to the interaction of light hydrocarbons with α2-u-globulin, widely recognised as confined to the male rat. Peroxisome proliferation is known to occur to a much greater extent in rodents than in man, if man exhibits it at all. There is a plethora of similar inconsistencies which often relate to toxicological sequelae of high doses resulting in prolonged/ repeated organ damage. Such responses serve to underline the importance of appropriate dose selection for studies to be used in risk assessment as well as the need for a knowledge of mechanism of toxicity.

This paper has attempted to illustrate some of the strengths and weaknesses of the chronic toxicity assessment. Even with internationally agreed protocols, there are sources of quantitative variation between laboratories which generally lead to the definition of no effect levels which are conservative. The studies can be useful qualitative predictors for effects in man but unusual changes may require supplementary information, particularly when these are seen mainly at high dose levels. Appropriate statistical evaluation may aid the interpretation of results obtained but ultimately their applicability for human risk assessment is a matter of judgement, (Paynter et al, 1985); the difficulty of regulating this commodity is readily acknowledged.

REFERENCES

Farber T.M. (1987) Selection of a Maximum Tolerated Dose (MTD) in Oncogenicity Studies: A position document of the US Environmental Protection Agency Office of Pesticide Programs EPA 540/09-88-003, Washington DC.

GIFAP (1982) Position paper: Chronic Toxicity in the Dog, GIFAP bulletin 8(4).

de Groot A.P. Feron V.J. and Immel H.R. (1988) Induction of Hyperplasia in the Bladder Epithelium of Rats by a Dietary Excess of Acid or Base: Implications for Toxicity/Carcinogenicity Testing. Fd. Chem. Tox. 26 425-434.

Hart R.W. and Chou M.W. (1991) Calorie Restrictions and Chemical Toxicity/Carcinogenesis, Proceedings of the First International Conference on Food, Nutrition and Chemical Toxicity, Surrey, UK (in press).

Haseman J.K. (1985) Issues in Carcinogenicity Testing: Dose selection. Fund. Appl. Toxicol 5 66-78.

Lumley C.E. and Walker S.R. (1986) A Critical Appraisal of the Duration of Chronic Animal Toxicity Studies. Reg. Tox. and Pharmacol 6 66-72.

Paynter O.E., Harris J.E., Burin G.J. and Jaeger R.B. (1985) Hazard Evaluation Division Standard Evaluation Procedure. Toxicity Potential: Guideline for Analysis and Evaluation of Sub-chronic and ChronicExposure Studies. EPA-540/9 85-020, United States Environmental Protection Agency, Washington DC.

TOXICOLOGICAL EVALUATION OF DIMETHOMORPH: BASIS FOR ASSESSMENT OF RISKS FROM DIETARY EXPOSURE.

G.E. Veenstra and D.E. Owen.
Shell Internationale Petroleum Maatschappij B.V., Health, Safety and Environment Division, P.O. Box 162, 2501 AN The Hague, The Netherlands.

INTRODUCTION.

Dimethomorph, is the common name for (E,Z) 4-[3-(4-chlorophenyl)-3-(3,4-dimethoxyphenyl) acryloyl]morpholine (Fig. 1). The material consists of a mixture of 2 geometric isomers (E and Z) in approximately equal amounts. The material is currently under development as an agricultural fungicide for use on potatoes and vines.

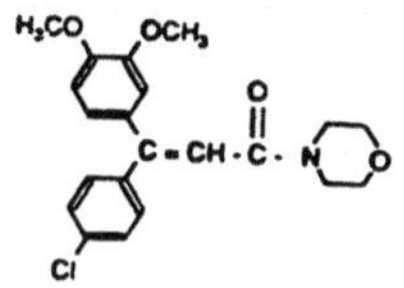

Figure 1.

A complete toxicological data base has been developed for dimethomorph to provide the basis for an assessment of the risks from chronic exposure through the intake of food containing residual amounts of dimethomorph.
In this paper, the relevant toxicological studies are summarised to provide the justification of the selection of the reference dose (RfD).

SUMMARY OF TOXICITY STUDIES.

Acute Toxicity.

Dimethomorph is of a low order of acute oral, dermal and inhalational toxicity. Oral and dermal LD50 values in the rat were 3900 and >5000 mg/kg, respectively. The oral LD50 value of the individual E and Z isomers was quantitatively similar to that of the mixture. The intra-peritoneal (i.p.) LD50 value in the rat was 321 mg/kg. The 4-hour inhalational LC 50 value was >4.2 mg/ml.
Dimethomorph was non-irritant to skin and eyes. Skin sensitisation potential was neither identified in an epicutaneous test nor in a guinea pig maximisation procedure.

Sub-chronic and chronic toxicity.

The cumulative toxicity of dimethomorph was evaluated in short-term (4-13 weeks) and long-term (52-104 weeks) toxicity studies in the rat, mouse and dog. Evaluation of the results showed the rat to be the most sensitive species for the toxic effects of dimethomorph. The studies of relevance for the determination of the reference dose are presented below.

In Sprague-Dawley rats, when tested at dietary dose levels of 40, 200, 1000, 2000, 3000 or 4000 ppm for periods up to 13 weeks, dimethomorph caused reductions in body weight gain and food consumption at doses of 2000 ppm and higher, with the effects in the females being more severe than in the males. There was a dose-related increase in relative liver weight in males at dose levels from 2000 ppm and in females at 1000 ppm and higher. Hepatocellular hypertrophy was noted at dose levels of 2000 ppm and higher, however, no degenerative hepatic lesions were recorded. In rats receiving dietary doses of 2000 ppm or more, ileal dilatation was observed. This finding was associated with chronic ileal inflammation originating from the serosal surface and was more severe in females than in males.

The toxicological profile of the E and Z isomers of dimethomorph in the rat was qualitatively identical. Oral administration by gavage of the E or Z isomer at daily doses of 0, 10, 100 or 750 mg/kg body weight/day to Fischer 344 rats for 28 consecutive days resulted in minimal effects on body weight, food consumption, clinical observations and haematology. Adjusted liver weights were increased at the 100 and 750 mg/kg/day level. Histological changes were limited to the liver which showed slight lipid vacuolation in the 100 and 750 mg/kg/day dose groups. Although the doses of the individual isomers in these gavage studies were considerably higher than in the dietary studies, inflammatory changes in the gastro-intestinal tract, as recorded with the mixture of isomers, were not observed.

The chronic toxicity of dimethomorph was evaluated in Sprague-Dawley rats (20 animals/group/sex) at dietary doses of 0, 200, 750 or 2000 ppm for 104 weeks. During the study a progressive loss in body weight gain developed in males at 2000 ppm and in females at 750 and 2000 ppm; the reduction in body weight gain at the 2000 ppm level exceeded 30% by the end of the study. Food consumption, however, was not affected by treatment. A dose-related increased survival was noted in males, with mortality in controls (55%) being more than twice that of the top dose animals (25%) after 104 weeks. There was a mild anaemia with a compensatory response in the bone marrow at 2000 ppm. Microscopic changes were limited to the top dose level and consisted of an increase in the incidence of hepatic 'ground-glass' foci and, in females, hepatocellular hypertrophy and brown pigment (lipofuscin), and a reduced incidence of proliferative findings in the pituitary. In males there was an increased incidence of mesenteric arteritis and testicular interstitial cell (IC) adenoma (statistically not significant). The changes in the tumour incidence of benign pituitary and testicular tumours observed in this study are consistent with the profile of spontaneous neoplasia under conditions of reduced body weight gain (Coneybeare, 1988).

Carcinogenicity.

The carcinogenic potential of dimethomorph was evaluated in the mouse and the rat. No evidence of a carcinogenic reponse was found in a carcinogenicity study in mice at dietary doses up to 1000 mg/kg body weight/day.

In a carcinogenicity study in Sprague-Dawley rats (50 animals/group/sex) dimethomorph was administered at dietary dose levels of 0, 200, 750 or

2000 ppm for a period of 2 years. Body weight gain and/or food intake were adversely affected in females receiving 750 or 2000 ppm; male body weight gain was decreased moderately at 2000 ppm. Histopathological examination showed changes limited to the top dose level and consisted of an increased incidence of hepatic 'ground-glass' foci, hepatocellular hypertrophy and pigmentation (lipofuscin), and, in males, an increased incidence of testicular IC adenoma (incidence 10 and 20 % for controls and 2000 ppm, respectively). The incidence was not statistically increased in a pair-wise comparison or a trend test. The range of the spontaneous incidence in 8 comparable studies was 4 - 20 %, with a mean incidence of 11.6 %. In the light of the range of the highly variable and spontaneous occurrence of testicular interstitial cell adenoma and the lack of any statistical significance within this study, it is concluded that dimethomorph is not a carcinogen in rats.

Reproductive and developmental toxicity

Dimethomorph was administered to Sprague-Dawley rats at dietary concentrations of 0, 100, 300 or 1000 ppm during all stages of a reproduction study for 2 generations. The first generation bred once, the second generation produced two litters. A decreased body weight gain was noted in females receiving 1000 ppm; at 300 ppm there was an equivocal effect on body weight. In males no adverse effects were observed. The fertility and reproductive performance, even at doses producing maternal toxicity, were not affected in any of the generations.

In teratogenicity studies, dimethomorph was administered to the rat and the rabbit during the period of organigenisis at daily doses up to 160 mg/kg/day for the rat and up to 650 mg/kg/day for the rabbit. In these studies, dimethomorph did not exhibit a teratogenic potential, even at doses eliciting marked maternal toxicity. There was some evidence of embryotoxicity (abortions and post-implantation losses) at doses which induced significant maternal toxicity; the embryotoxic effects were considered to be secondary to the maternal toxicity. No effects were noted on the foetuses at any of the doses or species tested.

Mutagenicity

The mutagenic potential of dimethomorph was evaluated in a battery of in-vitro and in-vivo experiments covering all major genetic end-points.

Gene mutation tests in microbial and mammalian cell test systems showed no increased mutation frequencies. A chromosome aberration test in V79 cells showed an equivocal response at 7 hours in the presence of a metabolic activation system. A cell transformation assay and a UDS test revealed no genotoxic activity. The significance of the results in-vitro was evaluated in a mouse micronucleous test, which after oral or i.p. administration showed an unequivocal lack of clastogenic activity. Based on the overall weight of the evidence it is concluded that dimethomorph is not a mutagen.

TOXICOLOGICAL EVALUATION

An evaluation of the results of the toxicological studies demonstrates that dimethomorph has a low potential for cumulative toxicity in the rat, mouse and dog. A common feature of the short term studies in all species tested is an increased functional hepatic demand; there is no evidence of degenerative hepatic effects.

The lowest no-observed effect level (NOEL) for chronic effects is 200 ppm in the rat (equivalent to 10 mg/kg body weight/ day). At the dose above the NOEL, 750 ppm, depression of body weight gain was observed in females. Dimethomorph was not carcinogenic in the rat and the mouse. The testicular effects in male rats of the chronic toxicity study at 2000 ppm represent a common finding in rats. The development and maintenance of testicular interstitial cell tumours is under hormonal control and of non-genotoxic origin (Chatani et al., 1990). Also, an increased incidence of testicular IC adenoma is associated with decreased body weight gain and is considered to be insignificant for human risk assessment. Based on the results of a battery of mutagenicity tests it is concluded that dimethomorph is not a mutagen. Developmental toxicity studies in rats and rabbits have not elicited any teratogenic or foetotoxic effects. The lowest NOEL for maternal and developmental effects is 60 mg/kg/day in the rat. In a 2-generation reproduction study in the rat, no adverse effects on reproduction were noted at dietary doses of 1000 ppm (50 mg/kg/day).

It is concluded that in the toxicological data base, covering studies to determine chronic, developmental and reproductive effects, no toxicological findings have been identified which are rare, unique or irreversible. In the absence of significant toxicological hazards, the use of an uncertainty (safety) factor of 100 to establish the RfD is justified. The lowest NOEL for chronic effects, 10 mg/kg/day, from the rat long-term studies represents an appropriate basis for the RfD of dimethomorph.

The proposed reference dose is 0.1 mg/kg/day.

REFERENCES

Chatani F, Nonoyama T, et al. (1990) Stimulatory effect of luteinizing hormone on the development and maintenance of 5α-reduced steroid-producing testicular interstitial cell tumors in Fischer 344 rats. Anticancer Res. 10: 337-342.

Coneybeare G. (1988) Modulating factors: Challenges to experimental design. In Grice HC and Ciminera JL (eds) Carcinogenicity, ILSI Monographs, Springer, New York, Berlin, Heidelberg, London, Paris, Tokyo, 149.

CHRONIC EXPOSURE TO VANADATE AS FACTOR OF ARTERIAL HYPERTENSION IN THE RAT: TOXICODYNAMIC MECHANISMS

M Carmignani*, AR Volpe*, G Porcelli*, P Boscolo**, P Preziosi***
*Dept. of Cell Biology and Physiology, University of L'Aquila, I-67010 Coppito, Italy
**Chair of Occupational Medicine, University of Chieti, I-66100 Chieti, Italy
***Inst. of Pharmacology, Catholic University School of Medicine, I-00168 Rome, Italy

Vanadium (mostly as vanadyl) was found to induce arterial hypertension by acting on peptidergic and catecholaminergic systems as well as on Ca^{2+} homeostasis in vascular and cardiac myocells.

1. Introduction

It is controversial whether cardiovascular (CV) effects of vanadate (V) and vanadyl involve inhibition and/or stimulation of enzymatic activities (like Na,K-ATPase and adenylate cyclase,respectively) and are influenced by other ions (like Na^{+},K^{+},Mg^{2+} and Mn^{2+}) and/or by the redox and metabolic state of the involved cells (Carmignani 1990). It is also uncertain whether the increase of cell Ca^{2+}, responsible of some effects of V such as vasoconstriction and inhibition of renin secretion, is due to a decrease in Ca^{2+} efflux (following inhibition of Ca-ATPase and Na, K-ATPase) or to an increase of Ca^{2+} influx through Ca channels (following inhibition of electrogenic Na,K transport) (Churchill 1990). This research investigated the possibility that chronic exposure to V leads to arterial hypertension and affects some effector, peptidergic and catecholaminergic mechanisms involved in CV regulation.

2. Material and Methods

Thirty male Sprague-Dawley rats (10 in each group) received 0 (control), 10 or 40 μg/ml of V (as $NaVO_3$) in drinking water for seven months starting from weaning. At the end of the treatment the activities of kallikrein (Morita 1977), kininases I (K1) and II (K2), and enkephalinase (ENase) (Volpe 1989) were determined in the 24-hour urines, in which creatinine, total nitrogen, proteins and electrolytes were also determined. Plasma renin activity (PRA) and levels of plasma aldosterone were evaluated in samples of arterial blood (Churchill 1990). Aortic blood pressure (BP), heart rate (HR) and maximum rate of rise of the left ventricular isovolumetric pressure (dP/dt) were monitored polygraphically under barbiturate anesthesia (Carmignani 1991). The following substances (μg/kg)were then injected by i.v. route:bradykinin (BK;1,11.5),ile-ser-BK (TK;11.5), leu^{5}- and met^{5}-enkephalins (LE,ME;100),angiotensin I (A1;0.50, 1) and II (A2;0.50), phenylephrine (PE;10), histamine (HIS;10), serotonin (SER;10), isoprenaline (ISO; 0.25),papaverine (PAP;2000),dibutyryl-cyclic AMP (dcAMP;5000),veratrine (VE;10), verapamil (VER;100) and hexamethonium (HEX;625, 1250).

The same substances (with the exception of TK, LE, ME and PAP) were given intravenously, at the same doses and before and after vagotomy, to twelve male Sprague-Dawley rats (6 in each group) exposed for seven months to 0 (control) and 100 μg/ml of V (as $NaVO_3$) in drinking water. Acetylcholine (Ach;2.50),tyramine (TYR;250), ouabain (OUA;20) and clonidine (CLO;10) were also assayed.

Samples of tissues from V-treated rats were submitted to instrumental neutron activation analysis for determining levels of vanadium in the pentavalent (V) and tetravalent (vanadyl) oxidation states (Carmignani 1990, 1991).

Data were expressed as means ± S.E.M. and compared by analysis of variance.Only a p value less than 0.05 was considered to be significant.

3. Results

BP (systolic, diastolic and mean), but not dP/dt, was increased in the groups of V-exposed rats in a similar manner; HR was augmented only in the rats exposed to 100 ppm of V (Tables 1,2). CV responses (CR) to A1, A2, TK, PE, HIS, SER, PAP and Ach were unchanged by V. On the other hand, in the rats treated with 10 and 40 ppm of V, the effects of BK,LE,ME, ISO and dcAMP on BP and dP/dt were potentiated by the metal, mainly at the lower dose, while those of VE,VER and HEX were not modified (Table 1). At

Table 1. Cardiovascular responses to several drugs, given by i.v. injection, in rats exposed for seven months to 10 ppm or 40 ppm of vanadate

Drugs (μg/kg)	Mean blood pressure (mmHg)			dP/dt ($mmHg/sec^{-1}x10^{-1}$)		
	Control	10ppm	40ppm	Control	10ppm	40ppm
Bradykinin (11.5)	-19±2	-43±3*	-35±3*	+218±17	+294±18*	+372±22*
Leu⁵-enkephalin (100)	-10±2	-30±5*	-18±6	- 86± 1	-166±17*	-139± 9*
Met⁵-enkephalin (100)	-19±4	-43±2*	-19±3	-102± 8	-182±13*	-171±16*
Isoprenaline (0.25)	-22±2	-37±2*	-29±5	+283±17	+449±32*	+371±24*
Dibutyryl-cyclic AMP (5000)	-17±4	-31±2*	-35±4*	+118± 8	+107±15	+171±12*
BASAL VALUES	92±5	114±3*	113±2*	421±13	398±12	440±13

Values are means ± S.E.M. (n=8 in each group)
*Significantly different from the control ($p<0.05$)

the dose of 100 ppm, V increased inotropic responses to BK, potentiated CR to ISO and reduced those to VE, increased BP responses to dcAMP, TYR, VER, HEX and CLO, reduced the effects of vagotomy on BP and HR and opposed those of OUA and CLO on dP/dt (Table 2).

In the rats exposed to 10 and 40 ppm of V, the activity of K1 was increased in a dose-related manner; the activities of kallikrein, K1, K2 and ENase were also significantly increased (Table 3). PRA (ng/ml/h) was 1.0±0.3, 4.7±1.5 ($p<0.05$) and 4.1±2.0 ($p<0.05$) in the rats treated with 0, 10 and 40 ppm of V, in which levels of plasma aldosterone (pg/ml) were

Table 2. Cardiovascular responses to several drugs, given by i.v. injection, before and after vagotomy in rats exposed for seven months to 100 ppm of vanadate

Drugs (μg/kg)	Mean blood pressure (mmHg)		Heart rate (beats/min)		dP/dt (mmHg/sec^{-1}x10^{-1})	
	Control	100ppm	Control	100ppm	Control	100ppm
Bradykinin (1)	- 9±2	-16±3	-12±3	- 9±3	+187±21	+321±18*
Isoprenaline (0.25)	-12±1	-22±3*	+17±5	+39±6*	+562±43	+1094±98*
Dibutyryl-cyclic AMP(5000)	-18±1	-29±2*	+31±5	+23±7	+588±53	+514±33
Veratrine (10)	-16±5	- 6±1*	-20±4	+ 6±2*	-160±17	-805±13*
Tyramine (250)	+20±6	+37±5*	+34±1	+29±6	-348±27	-374±43
VAGOTOMY	+18±3	+ 9±1*	+28±4	+ 5±1*	+107±71	+128±21
Ouabain (20)	+14±2	+18±4	-13±2	-10±4	+548±32	+231±68*
Verapamil (100)	-15±4	-35±5*	-40±4	-35±8	-428±20	-426±68
Hexamethonium (625)	-21±3	-39±3*	-12±3	-15±4	-294±27	-257±32
Clonidine[a] (10)	+15±2	+29±3*	-33±5	-35±3	-535±58	-342±11*
BASAL VALUES	104±3	125±5*	239±8	288±6*	877±79	852±50

Values are means ± S.E.M. (n=6 in both groups)
*Significantly different from the control ($p<0.05$)
[a] The initial hypertensive response (followed by long-lasting hypotension) was dependent on stimulation of the peripheral α_2-adrenoreceptors

Table 3. Activities of kallikrein, kininases I and II, and enkephalinase in the 24-hour urines of rats exposed for seven months to 10 ppm or 40 ppm of vanadate

Enzymatic activities	Control (n=6)	Exposed	
		10ppm (n=7)	40ppm (n=9)
Kallikrein[a]	8.43±0.96	13.67± 2.54*	11.72±0.80*
Kininase I[b]	32.02±4.22	129.93±14.96*	156.87±9.10*
Kininase II[b]	1.83±0.26	2.63± 0.13*	3.92±4.08*
Enkephalinase[b]	46.94±6.53	97.24±11.28*	73.81±5.66*

Values are means ± S.E.M.
*Significantly different from the control ($p<0.05$)
[a] The activity was expressed as nMoles/mg of 24-hour urine creatinine
[b] The activities were expressed as μMoles x10^{-3} of hydrolysed substrate/mg of 24-hour urine creatinine

188±57, 554±160 ($p<0.05$) and 265±61, respectively (5 in each group). In the same rats the urinary levels of creatinine, total nitrogen, proteins, sodium and calcium were not changed by V; however, the levels of urinary potassium (mEq/g creatinine) were 118±14, 169±18 and 221±28 ($p<0.05$) in the rats exposed to 0, 10 and 40 ppm of V,respectively (6 in each group). Vanadium was accumulated mostly in the kidney (within spheroidal electrondense structures in the cytoplasm of the proximal tubule cells) and in the bone; the aorta contained much more vanadium than the heart; vanadium was in prevalent form of vanadyl (Carmignani 1990, 1991).

4. Discussion

Chronic exposure to V was found to cause arterial hypertension depending mostly on increase of the total peripheral resistance. At lower level of exposure, V increases CR to BK, LE and ME (by acting on the renal kallikrein-kinin, enkephalin and renin-angiotensin-aldosterone systems) and to β_1- and β_2-adrenoreceptor agonists (by interfering with the cyclic-AMP-dependent availability of Ca^{2+} for contractile processes of both vessels and myocardium). At higher levels of exposure, V also reduces baroreflex sensitivity and vagal activity, enhances central sympathetic tone, potentiates CR to α_2-adrenoreceptor agonists (by probably increasing the verapamil-sensible influx of Ca^{2+} through the receptor-operated channels of vascular myocells), and stimulates the tyramine-sensible release of noradrenaline from the postganglionic sympathetic endings (Carmignani 1990, 1991). Moreover,V enhances the BK_2-receptor-dependent positive inotropic responses to BK, likely because of the increased activity of K1. The V-induced lower inotropic response to OUA seems to exclude that the in vivo cardiac effects of the metal depend only on binding of Na,K-ATPase (Carmignani 1991).

Vanadium (mostly as vanadyl) may influence the development of arterial hypertension through effects on catecholaminergic and peptidergic systems as well as on Ca^{2+} homeostasis in vascular and cardiac myocells).

Acknowledgements This work was supported by grants of Italian CNR nr. 88.01336.59 and MPI (M.C.) 40% and 60%.

5. References

Carmignani M, Sabbioni E, Boscolo P, Pietra R, Ripanti G (1990) Cardiovascular effects of long-term exposure to vanadate in rats. In: Anke M, Baumann W (eds) 6th international trace element symposium 1989 (Mo,V), vol 1. Karl-Marx-University Press, Leipzig, p 106

Carmignani M, Boscolo P, Volpe AR, Togna G, Preziosi P (1991) Cardiovascular system and kidney as specific targets of chronic exposure to vanadate in the rat: functional and morphological findings. Arch Toxicol Suppl 14: 124-127

Churchill PC, Rossi NF, Churchill MC, Ellis VR (1990) Vanadate-induced inhibition of renin secretion is unrelated to inhibition of Na,K-ATPase activity. Life Sci 46: 1953-1959

Morita T, Kato H, Iwanaga S, Takeda K, Kimura T, Sakakibara S (1977) New fluorogenic substrates for α-thrombin, factor Xa, kallikreins and urokinase. J Biochem 82: 1495-1498

Volpe AR, Porcelli G, Di Iorio M (1989) Determination of kininase I and II activities in human urine by HPLC. J Chromatogr 414: 427-431

Organ Toxicity

(Chair: G. H. Pigott, Macclesfield)

DERMAL TOXICOLOGY

Per E J Kulling

Swedish Poison Information Centre, Box 60500, S-104 01 Stockholm, Sweden

Dermal toxicology is a huge topic including dermal reactions after acute and chronic exposure. This paper, however, will mainly focus on non-immunological dermal reactions after acute exposure. Dermal effects might be due to local injury after skin exposure or may be signs of systemic poisoning regardless route of exposure or a combination of both.

The skin consists of three main layers: the epidermis, the dermis (corium) and the subcutis. The outer layer, the horny layer (stratum corneum), consists of several layers of flattened, keratinized, dead cells. The deeper layers of the epidermis consist of cell layers that exhibit increasing degrees of flattening and keratinization toward the stratum corneum. Below the stratum corneum, the dermis forms papillae that contain blood vessles.

Healthy skin is a very good barrier which limits both loss of water and other essential compounds from the body and diminishes the penetration of potentially toxic substances into the body.

SKIN EXPOSURE

In a survey of enquiries to the Swedish Poison Information Centre during one year (1 February 1987 - 31 January 1988) concerning acute occupational exposure to chemicals skin exposure occured in 22 % of the 883 cases included.

Skin exposure to toxic agents can give rise to local injury, but systemic absorption of toxic substances might also occur resulting in systemic poisoning. Mainly lipid-soluble substances undergo percutaneous absorption while water-soluble substances do not penetrate the skin (Grandjean 1990). However, if the skin is damaged in one way or another the barrier function is destroyed and substances that normally do not penetrate the skin might be absorbed. A man was for example splashed with a concentrated solution of monochloroacetic acid, a water soluble substance that normally does not penetrate skin as such. He developed first and second degree corrosive burns over exposed skin areas. Within a couple of hours he developed signs of severe systemic poisoning (cardiodepression, seizures, metabolic acidosis, renal insufficiency) indicating absorption through the skin, which also was verified by blood analyses (Kulling et al).

Chemicals may be more toxic when absorbed through the skin than if absorbed via the oral route, as detoxification in the liver may be bypassed. On the other hand, metabolizing enzymes in the skin may also affect the system toxicity of chemicals absorbed by this route. The skin contains many of the enzymes that also occur in the liver. A chemical can in this way either decrease or increase its systemic toxicity (Grandjean 1990).

This paper will focus on dermal toxicology while aspects on skin absorption and concomitant systemic poisoning only will be given when appropiate.

Local Injury

Local injury might affect all layers of the skin. When chemical substances penetrate the skin local toxicity may occur in the different layers of the skin.

Agents causing local injury are chemicals (including drugs), plants and animals. The type of acute injury from these different agents can be classified as follows according to their action: allergic action, irritant action, corrosive action, cytotoxic action or combinations of these. The definitions of these groups are vague and a number of circumstances is of importance for example whether the action should be regarded as an irritation or a corrosion. The local allergic action can express itself as a minor redness and pruritus or a most dramatic reaction including severe edema and skin damage. An allergic reaction is dose independent.

Factors of importance for severity: A number of factors of importance for severity of injury in non-allergic reactions can be identified. The amount of the toxic agent and the concentration of a solution are of great importance in this respect. Other factors of importance are duration of exposure and if exposure is under pressure. Gasoline for example is regarded as a mild irritant, but prolonged exposure can give severe irritation or rather corrosivelike reaction with blisters. A few examples will illustrate this. A boy fell into a pool of gasoline and was lying there for some time. He developed blisters on those parts being exposed most intensely. A man developed severe blisters on his thigh after having had a wad steep with gasoline for a number of hours in his pocket (Swedish Poison Information Centre, unpublished data).

Interaction of substances may also be of importance. For example the combination of a strong acid and some water will result in an extensive exothermic reaction superimposing a thermal skin burn. A substance that normally only is an irritant can when combined with another substance become corrosive. For example skin exposure to a combination of paraffin and methylene chloride might give blisters, while neither of these two substances in themselves will do so (Swedish Poison Information centre, unpublished data).

A vital step in limiting tissue damage after skin exposure is first-aid. Different first-aid measures will be discussed later.

Mechanisms of non-allergic local injury: A number of different mechanisms of the non-allergic local injuries after skin exposure can be identified. The definition of the different mechanisms is vague and overlap many times. The attempt to classify the mechanisms here are rather based on a practical approach than on a strict scientific approach.

The most common mechanism and also the less strictly defined one is irritation. There is a great number of agents causing irritation. The definition of dermal irritation is the production of reversible inflammatory changes in the skin following the application of a test substance (OECD Guideline 1981). The biochemical mechanism involved in irritation is not well characterized. A substance might under certain circumstances just give irritation while it under other circumstances gives corrosion as discribed above (factors of importance for severity of damage).

The local skin injury seen after exposure to many petroleum products and other hydrocarbons (including many solvents) is due to the defattening properties of these agents. The lesion will in most instances just be irritation, but after prolonged exposure the damage might be more like a corrosion.

Corrosion is according to the definition the production of irreversible tissue damage in the skin following the application of a test substance (OECD Guideline 1981). Corrosion is manifested by ulceration and necrosis with subsequent scar formation. More severe corrosive changes are often designated as chemical burns. Certain substances are many times corrosive in higher concentrations while in lower concentrations they are just irritants.

The damage is in nature very much the same as seen in thermal burns and may be classified accordingly: epidermal injury, 1st degree burn, involves the superficial layer of the skin; dermal injury, 2nd degree burn, involves the dermis but heal without scar formation; full-thickness injury, 3rd degree burn, involves all layers of the skin and heals with scar formation. Following definitions strictly only full-thickness injury should be classified as a corrosive damage, and superficial and dermal injury should be classified as irritation.

Complications from corrosive injuries are the same as for thermal burns i.e. intense pain, fluid loss, and later risk of infections. Fluid loss can in severe cases be excessive leading to circulatory and other complications.

Characteristics of the lesions vary depending on the nature of the toxic material. Severe burns from acids usually cause a dry crust from coagulation necrosis. The coagulation necrosis stops the acid from further penetration into the skin. Alkalies like potassium and sodium hydroxide and ammonia produce softer burns and have a tendency to penetrate deeper as long as exposure occurs. They combine with proteins and lipids resulting in liquefaction necrosis thus penetrating to deeper layers. Many aliphatic amines act in a similar way as alkalies but are "weaker". Phenol denaturates proteins and produce chemical burns characterized by a whitish or brownish surface. Initially the skin becomes painless. There is also a risk of systemic poisoning after skin absorption of phenol.

The toxic action from some peroxides and chromates is attributed to the oxidizing properties of these agents. The type of damage is corrosive, but it does not penetrate to deeper layers as alkalies do.

Tissue fixation is seen after exposure to for example formaldehyde. This damage is a coagualtive necrosis producing a tanned leather-like lesion.

White or yellow phosphorus will self-ignite in air thus giving rather a thermal burn.

In situations when an exothermic reaction occurs as is seen for example when a strong acid or quicklime (CaO) comes in contact with moist skin, the heat production will superimpose a thermal burn on the corrosive injury.

A more "specific" cytotoxic action can also be identified in this attempt to classify mechanisms of local skin injury after skin exposure. Alkylating agents like mustard gases will give a local skin injury very much similar to a corrosive burn. The development of symptoms after skin exposure to alkylating agents is, however, delayed for a number of hours. The skin lesion is characterized by blisters containing a yellowish fluid. Severe cases heal with scar formation. The severe cases also show signs of systemic poisoning due to absorption of the toxic agent.

Strong solutions of hydrofluoric acid give immediate symptoms as for acids. However, the fluoride ion has a tendency to penetrate the skin to deeper layers reacting with tissue calcium and magnesium cations to form insoluble salts (calcium salt precipitation). This is thought to interfere with cellular metabolism, resulting in cellular death and necrosis. It is believed that potassium ions are also released and contribute to the extensive pain as is seen after skin exposure to hydrofluoric acid (Mac Kinnon 1988). Solutions stronger than 50 % will give symptoms within one hour, solutions of 20-50 % will give symptoms after 1-8 hours while symptoms after exposure to solutions of < 20 % may be delayed for up to 24 hours (Mac Kinnon 1988). Apart from the local injury there is also a risk of severe systemic hypocalcemia after skin exposure to hydrofluoric acid.

Exposure to for example creosote and the giant hogweed (Heracleum species) induces the skin to a phototoxic reaction when exposed to the ultraviolet component of sunshine. Swelling and redness frequently occur, and extensive blistering may be seen. The reaction is a non immunologic reaction including an increased reactivity of the target tissue to ultraviolet radiation (Emmet 1986). The reaction is dose dependant both concerning exposure to the toxic agent and to exposure to ultraviolet light.

After skin exposure to living animals either by stings or bites local skin reaction is mostly attributed to enzymatic activity by the injected toxin. For example the European adder venom toxin contains proteases and hyaluronidase (Harris 1982). The local reaction can vary from mild redness and minor swelling to extensive swelling, discolouration of the skin from local hemolysis, and necrosis. In severe cases the skin reaction might be a combination of toxic reaction and local tissue ischemia. Skin necrosis can be so extensive that skin transplantation can become necessary. Skin reaction and signs of systemic poisoning are in these cases independant from each other.

First-aid: A vital step in limiting or avoiding tissue damage (and absorption) is rapid first-aid measures.

Flushing with copious amounts of water as soon as possible shall always be performed after exposure to toxic agents giving local tissue damage. Flushing should be prolonged thus avoiding damage in those situations heat production may occur or corrosive agents may be formed. Small amounts of water on a strong acid will result in an extensive exothermic reaction necessitating water to neutralize the heat produced. The hydrolizing of for example titanium tetrachloride or tin tetrachloride to hydrochloric acid also necessitates the splashing with plenty of water not to worsen skin damage. In these situations it is advisable that before flushing with water, the exposed area is wiped with a dry piece of cloth. Exposed parts of the body to white phosphorus (which self-ignites in air) should always be kept under water until proper washing and debriement is performed. There is never indication for neutalizing acids and alkalies after skin exposure by using a weak solution of an alkali or acid respectively.

Contaminated clothing, shoes, wrist watches, and jewellary should always be removed to facilitate proper washing and stopping further exposure. If there is a risk of skin absorption or the agent is insoluble in water the affected skin area should be washed thoroughly with soap and water.

In a limited number of cases local administration of antidotes is of great importance to reduce or diminish local and systemic toxicity.

Soluble calcium salts as solution or jel are used in hydrofluoric acid skin exposure. The calcium ion reacts with the fluoride ion forming an insoluble salt thus "neutralizing" the toxic fluoride ion. This therapeutic measure is very efficient and will, when given in time, avoid or dramatically reduce skin damage and systemic poisoning.

After exposure to phenol the use of polyethylene glycol 400 or glycerin is used as solvents. Phenol is poorly soluble in water.

A solution of potassium permanganate and sodium bicarbonate is used after exposure to white phosphorus. The phosphorus is converted to phosphoric acid by the potassium permanganate which is further neutralized by the sodium bicarbonate.

Dermal Reaction As A Sign Of Systemic Toxicity

A number of non-immunological skin reactions can be signs of systemic toxicity. Such reactions are erythema, hyperpigmentation, vitiligo, discolouration, alopecia, nail changes, phototoxicity, acne production and bullous skin eruption. Many of these changes are rather seen in chronic or subchronic exposures than in acute exposures. When seen after acute exposure they are delayed reactions not presented until after a number of days or even weeks.

Acne production is a sign that can develop some time after acute exposure to toxic agents. A special type of chemical acne is chloracne which is seen after poisoning by halogenated aromatic compounds with a specific molecular shape (Poland & Glover 1977). The most potent chloracnegen is TCDD (2,3,7,8-tetrachlorodibenzo-p-dioxin). Chloracne is characterized by small, straw coloured cysts and comedones. Inflammatory pustules and abscesses may occur (Emmet 1986).

Bullous skin eruptions are seen occationally in severe poisoning from some sedative hypnotics, narcotics and carbon monoxide (Ellenhorn & Barceloux 1988; Moshkowitz et al 1990). It has also been seen in cases of prolonged coma of other origin than poisoning. Sedative hypnotic drugs in which this phenomenon has been described are barbiturates, meprobamate, tricyclic antidepressants, glutethimide and some bensodiazepines. A number of mechanisms has been suggested for this bullous eruption formation: local pressure, trauma, local hypoxia or a direct toxic effect. Histological examinations have shown epidermal necrosis, intrademal vesiculation and sweat duct and gland necrosis. The latter could be secondary to excessive sweating leading to "fatigue" of the sweat glands and subsequent necrosis.

CONCLUSIONS

In conclusion one can state that skin changes migt be seen in acute poisoning both after skin exposure and as signs of systemic poisoning regardless route of exposure. The skin reactions can be very severe, even life-threatening. After skin exposure proper and rapid first-aid measures can prevent or dramatically diminish the skin damage (and systemic poisoning).

REFERENCES

Ellenhorn MJ & Barceloux DG (1988) Medical Toxicology Diagnosis and treatment of human poisoning. Elsevier, New York Amsterdam London

Emmett EA (1986) Toxic responses of the skin. In: Klaassen CD, Amdur MO, Doull J (eds) Casarett and Doull's Toxicology, 3rd edn. Macmillian Publishing Company, New York, p 412

Grandjean P (1990) Skin penetration: Hazardous chemicals at work. Taylor & Francis, London New York Philadelphia

Harris JB (1982) Toxic constituents of animal venoms and poisons. Adv Drug React Ac Pois Rev 1: 65-92

Kulling P, Andersson H, Boström K, Johansson L-Å, Lindström B, Nyström B (to be published) Lethal systemic poisoning after skin exposure to monochloroacetic acid. J Toxicol Clin Toxicol

Mac Kinnon MA (1988) Hydrofluoric acid burns. Dermatologic Clinics 6: 67-74

Moshkowitz M, Pines A, Finkelstein A, Hershkowitz R, Levo Y (1990) Skin blisters as a manifestation of oxazepam toxicity. J Toxicol Clin Toxicol 28: 383-386
OECD Guideline for testing of chemicals (1981) Acute dermal iritation/corrosion. No 404
Poland A, Glover E (1977) Chlorinated biphenyl induction of amyl hydrocarbon hydroxylase activity:a study of the structure activity relatonsships. Mol Pharmacol 13: 924-938

Further reading:

Ellenhorn MJ & Barceloux DG (1988) Medical Toxicology Diagnosis and treatment of human poisoning. Elsevier, New York Amsterdam London
Gosselin RE, Smith RP, Hodge HC (eds) (1984) Clinical Toxicology of Commercial Products, 5th edn. Williams and Wilkins, Baltimore London
Haddad CM, Winchester JF (eds) (1983) Clinical Management of Poisoning and Drug Overdose. WB Saunders Company, Philadelphia London Toronto Mexico City Rio de Janeiro Sydney Tokyo
Moeschlin S (1986) Klinik und Therapie der Vergiftungen, 7th edn. Georg Thieme Verlag, Stuttgart New York
Rumach B (ed) PoisIndex (1991). 69th edn. Micromedex Inc, Denver

CYCADS, DESIGNER DRUGS, FREE RADICALS AND NEURODEGENERATIVE DISEASE

J. D. MITCHELL

Department of Neurology, Royal Preston Hospital, Sharoe Green Lane, Fulwood, PRESTON, PR2 4HT, UK.

GENERAL INTRODUCTION

There is a vast potential for substances to exert a deleterious effect on the nervous system. Neurotoxins vary widely in their molecular complexity from the elemental level in the case of mercury and lead to complex macromolecules with a molecular weight of up to 150,000 as in the case of botulinum toxin, arguably the most toxic substance known to man. Mercurial encephalopathy was immortalised by the "Mad Hatter" in the Alice in Wonderland stories of Lewis Carroll. It has been said that this author derived inspiration for many of these writings after eating the so-called "magic mushrooms" which grew on the banks of the Cherwell near Oxford and it is thus quite possible that many of these stories were conceived under the influence of an hallucinogenic neurotoxin. The neurotoxic sequelae of lead poisoning were well known to the ancients and were described by the Greek poet-physician Nikander before the birth of Christ (Mitchell 1987).

The field of neurotoxicology therefore covers a very wide spectrum and it is unfortunately not possible to provide a comprehensive review of the whole area in the space available. It is more sensible to restrict the substance of this paper to aspects which are of particular concern to the clinical neurologist and which represent current growing points in this field. These mainly centre on the neurodegenerative diseases and their possible relationship to putative environmental agents. Of the three principle neurodegenerative disorders, this paper will be mainly concerned with only two, amyotrophic lateral sclerosis (ALS) and Parkinson's disease (PD).

AMYOTROPHIC LATERAL SCLEROSIS (ALS) AND LYTICO BODEG

ALS is one of the most devastating and mysterious diseases known to medical science. There are numerous anecdotal reports of ALS-like syndromes arising in relation to all sorts of agents and influences. Examples include the dietary consumption of lupin seeds sold as "Specialité Portuguese" in France, exposure to various solvents and pesticide aerosols (chlordane, permethrin, Freon). One patient developed the disease at a particularly early age having been in the habit of walking and running through fields of Castor Oil plant (Ricinus) with bare legs as a boy. The possible aggregation of cases within the San Francisco 49ers, an American football team has been publicised in Time magazine. A specific fertiliser had been applied to the field on which they played and it was suggested that this may have been a factor in the causation of the disease in these individuals (Mitchell - in press).

There has been considerable recent interest in an ALS variant which is endemic in many Pacific areas. This variant has been particularly studied on the Pacific Island of Guam where it is known to the native population as lytico bodeg. It exists there in two main clinical forms which also encompass dementia and extrapyramidal manifestations. The first form (which most resembles ALS) has declined in recent years. The second (which is characterised by dementia and extrapyramidal manifestations) has not. This form of the disease closely resembles the clinical picture of the Steele Richardson Syndrome or progressive supranuclear palsy (PSP) and begins to raise the possibility of commonality in relation to the major neurodegenerative disorders.

In recent years there has been renewed interest in Cycads in relation to lytico bodeg. The Cycad nut has formed a dietary staple for the indigenous population and has been found to contain the amino acid BMAA (β-N-methylamino-L-alanine), which is very similar to BOAA (β-N-oxalylamino-L-alanine), a related amino acid which is found in the chickling pea (Lathyrus sativa) and is implicated in the aetiology of Lathyrism (Spencer et al 1986). The link between lytico bodeg and BMAA remains unproven. A washing procedure to remove toxic materials is routinely undertaken by the indigenous population before the nut is prepared for consumption. This involves washing with water. BMAA is highly water soluble and may well be effectively removed from the nut by this procedure. Other evidence also suggests that any BMAA which might enter the intravascular space is likely to be largely excluded from the circulation by first pass metabolism in the liver. Also BMAA probably crosses the blood brain barrier relatively poorly in view of its solubility in water. It is nonetheless possible that some other constituent of the Cycad nut is implicated in lytico bodeg. Cycasin (α-amino-β-methylaminoproprionic acid) has been previously studied in this context. This substance is a known carcinogen and may well be worthy of renewed attention and study as well as its aglycone, methyl azoxymethanol (MAM). Whatever the cause of this illness it must be remembered that the nut of the Cycad is used in all areas where a lytico bodeg like illness is recognised. It has been employed in folk medicine in the Kii peninsula of Japan and as a wound dressing in the Irian region of New Guinea. It must therefore remain a strong possibility that something connected with the cycad is a factor in the causation of these illnesses.

PARKINSON'S DISEASE

Another area of intense interest and importance in recent years has been the recognition of PD associated with exposure to 1-methyl-4-phenyl-1,2,3,6-tetrahydropyridine (MPTP), (Langston et al 1982). Drug dependence is unfortunately a major problem in many parts of California where the scene was set some years ago for the "designer drug" epidemic. This sprang from efforts by clandestine chemists to synthesise molecules with chemical structures closely resembling those of substances known to have the pharmacological properties of opiates, with the aim that the resulting compounds would have similar pharmacological effects to this group of drugs but differ sufficiently to avoid statutory restrictions on the supply and use of such materials.

In 1982 a group of drug addicts from this area presented with a syndrome closely resembling PD. Two particular patients were identified who had used the same synthetic heroin. In 1979 a research chemist had developed a similar disorder while working elsewhere on a meperidine-like drug known as MPPP. In taking short cuts in his work producing this compound, it had become contaminated with the pyridine derivative, MPTP. This substance was also found to be a contaminant of the synthetic heroin used by the Californian drug addicts. The tale took on a curious twist when a 1947 reference to a method of synthesising MPTP was found. This had been carefully removed from its volume in the Stanford University Library. When a copy had been finally found it became clear that MPTP had been investigated as a possible _treatment_ for PD in the late 1950's.

Over 300 cases of MPTP PD have now been recognised. It is clear that its resemblance to idiopathic PD becomes closer the more it is studied. Some patients have now been followed over a 5 year period and have shown the continuing slow progression typical of idiopathic PD during this time despite being no longer exposed to the putative toxin. They respond to levodopa in the same way and some also show the on/off phenomenon typical of some PD patients. Man seems particularly sensitive to the neurotoxic effects of MPTP. Other primates may develop neurotoxicity following exposure, but the characteristic tremor of the human model is lacking in these species. Rats seem immune to the effects of MPTP. In this substance a material has therefore been found which is not only selectively toxic to a defined neuronal system in man, but also reproduces the clinical features of a recognised human neurodegenerative disorder.

The work of Langston's group has also resulted in an additional concept of crucial importance in this context, that is the idea that a toxic response may only become clinically manifest some years after the critical exposure to the toxin. This concept is based on the idea that an individual is born with a definite neuronal complement which declines naturally with increasing age. If this neuronal complement falls below a critical threshold the loss becomes clinically apparent. In the case of ALS the neuronal system in question would be motor neurones at all levels: in the case of PD the pigmented cells of the substantia nigra. Whether a particular individual becomes symptomatic clearly depends on both the rate of neuronal loss and longevity. It is possible that an individual might encounter a substance selectively toxic to such a neuronal system at a specific time during life. This might result in an acute reduction of the neuronal population at that time which is not sufficient to tip the individual over the critical threshold. This insult, together with the continuing natural loss associated with the aging process, might however result in this threshold being crossed some years later, earlier than if the toxic exposure had not occurred. This is a potentially crucial concept in our attempts to understand neurodegeneration in the context of environmental toxins (Langston 1990).

FUNDAMENTAL MECHANISMS OF NEURODEGENERATION

This chapter has so far been clinically oriented. It is important not to forget the importance of understanding the fundamental mechanisms of neurodegeneration in this context. This issue will be examined in the rest of this work. Although a wide range of putative toxins have been mentioned, it is quite possible that their toxic effect is mediated through only a relatively small number of mechanisms. Many of these substances can be associated with the formation of free radical species. These intermediates may represent a final common pathway of the neurodegenerative process. Although the nervous system is specially vulnerable to degenerative disease as well as to damage from free radical species, there has so far been comparatively little work seeking to link free radical mechanisms with neurological disorders.

What are free radicals?

Free radicals are defined as chemical species containing one or more unpaired electrons. Because of inherent instability their existence is often transitory. In spite of this they are able to initiate considerable damage which may eventually result in cell death. Free radicals are thus of great potential interest in relation to neurodegenerative disease.

The dangers posed by these species are such that specific defence mechanisms exist both within the extracellular and intracellular compartments. These protective functions are mainly enzymatic in nature, but vitamins E (tocopherol) and C (ascorbic acid) are also important antioxidants. The brain is particularly rich in oxidisable substrates. These are mostly catecholamines and unsaturated lipids. Transition metal ions are also critical in relation to free radical mechanisms. Because of their ability to exist in multiple oxidation states they may in some cases facilitate the formation of free radical species.

The two most important naturally occurring radical species are the superoxide and hydroxyl radicals. The superoxide radical is the one electron reduction product of molecular oxygen (1).

$$O_2 + e^- \longrightarrow \underset{\text{Superoxide radical}}{O_2^-} \text{ --------- } (1)$$

An important natural source of superoxide is the activity of the mitochondrial and microsomal electron transport chains. Superoxide is converted to hydrogen peroxide in aqueous solution by a process known as dismutation (2). This reaction is catalysed

$$2O_2^- + 2H^+ \xrightarrow[\text{Superoxide dismutase}]{} H_2O_2 \text{ ----------- } (2)$$

by dismutases which, interestingly, are all metalloenzymes. Copper, zinc and iron containing forms have been described. The most dangerous situation arises when the hydroxyl radical ($\cdot OH$) is formed from the splitting of a hydrogen peroxide molecule (3, 4). This can be triggered by transition metal ions. It will be

$$H_2O_2 + Fe^{2+} \longrightarrow Fe^{3+} + \cdot OH + OH^- \text{ -------- } (3)$$

$$H_2O_2 + Cu^{+} \longrightarrow Cu^{2+} + \cdot OH + OH^- \text{ -------- } (4)$$

noted that by participating in the reaction, the metal ion enters a higher oxidation state. Low molecular weight iron complexes (Fe citrate, Fe ATP) are probably the main catalysts of this reaction and are found, as well as low molecular weight copper complexes, in human cerebrospinal fluid (CSF). The brain also has a high endogenous iron content. The hydroxyl radical is a highly unpleasant species and is able to interact deleteriously with almost all biological molecules. The most dangerous effects are those on nucleic acids and membranes.

Defence mechanisms against free radicals

Ascorbic acid is an important antioxidant when present in high concentrations. High concentrations of ascorbic acid are found both in the grey and white matter of the brain and the choroid plexus has a specific transport mechanism which concentrates ascorbate in the CSF. Although some have reported CSF ascorbate levels to exceed serum content by a factor of up to 10 we have found that the CSF/plasma ratio is rather less, usually ranging form 3 - 6. We have also found a clear logarithmic correlation between CSF and plasma ascorbate levels (correlation coefficient=0.84).

In addition to superoxide dismutases (SOD) other enzymes are involved in defences against free radicals. Brain levels of catalase are low . The selenoenzyme glutathione peroxidase (GSHPX) is likely to be more important in the inactivation and breakdown of peroxide accumulating in the brain.

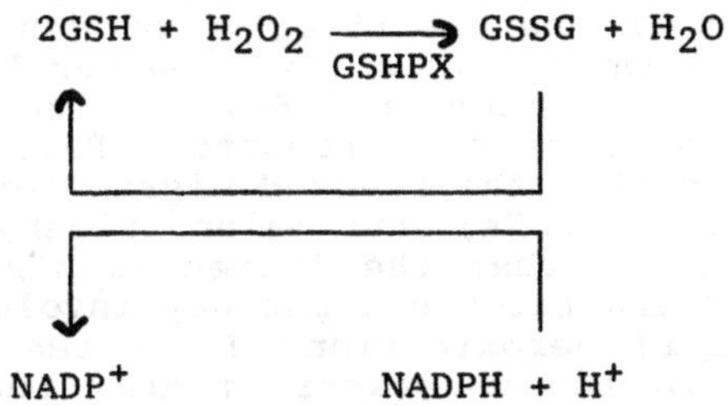

---------- (5)

This reaction hinges on the oxidation of reduced glutathione (GSH) to oxidised glutathione (GSSG). The reformation of GSH from GSSG is dependent on glutathione reductase, an enzyme also present in high concentrations in the brain (5).

Vitamin E (α-tocopherol) is also an important antioxidant. Because it is fat soluble it has a special role in preventing lipid peroxidation, a self perpetuating process which causes progressive damage to membranes if allowed to proceed unchecked. The neurological sequelae of vitamin E deficiency are well recognised in man. Patients with a-β-lipoproteinaemia suffer from malabsorptive tocopherol deficiency and develop a neurological syndrome which clinically resembles spinocerebellar degeneration (Halliwell & Gutteridge 1985a).

Extracellular mechanisms relating to the nervous system are poorly understood. Human CSF contains little catalase or transferrin and we have found very low levels of caeruloplasmin, SOD and GSHPX. We have found that vitamin E is virtually undetectable in CSF, a finding which is not particularly surprising in view of its lipid solubility. Low molecular weight iron and copper complexes are however present and it will be recalled that these can act as powerful promoters of free radical formation. It seems clear that CSF must contain an effective antioxidant. Although the precise nature of this substance is not yet known, it does seem possible that ascorbic acid may have an important role in this regard.

Free radicals and Parkinson's disease.

The capacity for catecholamines to produce free radical derivatives has been fundamental in ideas that the degenerative process in PD may at least partially depend on free radical mediated mechanisms. The neurological disease known as "locura manganica" has many features of PD and was related to occupational exposure in Chilean manganese miners in the classical work of Cotzias (Cotzias et al 1968). It has since been suggested that this disease may have resulted from formation of catecholamine free radical derivatives triggered by manganese

(Halliwell & Gutteridge 1985).

It must however be stressed that although similarities exist, locura manganica is clearly not the same disease as idiopathic PD. Studies of brain transition metal levels in PD have not shown any changes in the distribution of manganese. Increased iron and reduced copper levels have however been found in the substantia nigra in PD (Jenner 1989) and increased CSF copper concentrations have also been reported (Pall et al 1987). It has been suggested that the reduced nigral levels of polyunsaturated fatty acids (PUFA) and malondialdehyde levels in PD raise the possibility that the increased nigral iron content stimulates cell death through a pathway involving free radical formation and lipid peroxidation. Increased SOD levels have also been found in the substantia nigra. The evidence would suggest that this increase is due to the manganese containing, mitochondrial form of the enzyme. Recent work has suggested an impairment of complex I of the mitochondrial respiratory chain in PD (Jenner 1989).

Soon after the discovery of MPTP PD (Langston et al 1982) it was suggested that the toxic principle was not MPTP but its metabolite MPP^+. The charged nature of this moiety, as well as its structural similarity to paraquat lead to suggestions that the toxic effect might ultimately depend on free radical mechanisms. It now seems unlikely that MPP^+ is sufficiently unstable to result in the formation of toxic oxyradicals. The metabolism of MPTP itself has however been shown to be associated with radical formation. This has been found to occur in aerobic mitochondrial preparations and may be prevented by the addition of SOD. These comments are clearly of interest in relation to reports of an impairment of the mitochondrial respiratory chain in idiopathic PD (Jenner 1989).

Free radicals and ALS

Like PD, ALS is a systematised neuronal degeneration. Reduced RNA levels have been found in surviving motor neurones with preservation of DNA content (Davidson & Hartmann 1981). This has been considered to suggest a defect of DNA transcription in ALS. Increased selenium levels have been reported in erythrocytes (Nagata et al 1985), cervical cord, liver and bone (Mitchell et al 1991). Increased spinal cord (Miyata et al 1983, Mitchell et al 1991) and reduced hepatic (Mitchell 1991) manganese levels have also been found. Manganese can itself trigger free radical formation and is also a constituent of mitochondrial SOD. Selenium is a structural component of GSHPX. These findings have been interpreted as suggesting that free radical mediated mechanisms might be implicated in the initiation of the degenerative process in ALS. Free radical induced damage to polynucleotides might disrupt transcription. Increased spinal cord selenium and manganese levels may also indicate increased GSHPX and MnSOD activity.

We first tested this "free radical hypothesis" of ALS by examining CSF parameters of free radical activity and lipid peroxidation. Diene conjugates were measured as a reflection of the early stages of lipid peroxidation. Chloroform fluorescence was regarded as an index of the later stages of lipid peroxidation and methanol fluorescence as a measure of free radical damage to proteins. No differences in any of these parameters were found between ALS patients and controls (Mitchell et al 1987).

This result was not necessarily surprising. The blood brain barrier is intact in ALS and correlates of abnormal free radical activity might not be found in CSF. It is to some extent surprising that the results of these studies were so unremarkable. Some changes might even have been expected as an epiphenomenon. Also, the indices considered in this work mainly related to free-radical induced membrane damage. The evidence suggests that the fundamental trigger of ALS acts at an intracellular rather than a membranous level. In view of this we attempted to measure GSHPX activity in spinal cord and liver of patients dying of ALS. Although only a small number of results are so far available the results show a wide variation and some very low GSHPX activities have been found. In general, the activities seem higher in white than in grey matter. When an analysis comparing results from these samples is made, the GSHPX activities seem rather lower in the ALS patients than in the controls (Mitchell & Pentland 1990). This would however be difficult to reconcile with the high spinal cord, liver and bone selenium levels.

GENERAL CONCLUSIONS

Although the ultimate causes of the neurodegenerative diseases remain unknown, toxicological investigations are crucial in determining whether these disorders might be related to exposure to environmental toxins. Such studies must attempt to define the way in which such substances might initiate the neurodegenerative process. It is possible that a multiplicity of toxins are involved and genetic susceptibility may also be a factor. This may be one reason why this area currently seems so confusing. The neurodegenerative diseases remain one of the great mysteries of medical science and are the cause of untold human suffering. In the light of current advances, however, it is to be hoped that significant progress in this area is a realistic prospect as we approach the year 2000.

ACKNOWLEDGEMENTS

The author is indebted to the following for their help in the work described here. Dr. MJ Jackson of the Department of Medicine, University of Liverpool, Mr. JA Gatt and Miss G Rostron of the Department of Clinical Chemistry, Royal Preston Hospital, Dr. PG Lynch, Consultant Neuropathologist, Preston, Mrs. E Houghton, Department of Neurology, Royal Preston Hospital. Also Dr. B Pentland, Astley Ainslie Hospital, Edinburgh and Dr. BW East and Mr IA Harris of the Scottish Universities Research & Reactor Centre, East Kilbride.

This work has been supported by the Motor Neurone Disease Endowment Fund of the Preston Health Authority, George Barton Trust, Britannia Pharmaceuticals and the Preston & Chorley Hospitals Research Fund. The author is also supported by MNDA (UK).

REFERENCES

Cotzias GC, Horiuchi K, Fuenzalida S et al (1968) Chronic manganese poisoning. Clearance of tissue manganese concentrations with persistence of the neurological picture. Neurology 18: 376-82

Davidson TJ, Hartmann HA. (1981) RNA content and volume of motor neurones in amyotrophic lateral sclerosis. J Neuropath Exp Neurol 40: 187-92.

Halliwell B, Gutteridge JMC. (1985a) Oxygen radicals and the nervous system. Trends in Neurol Sci :22-26

Jenner P. (1989) Clues to the mechanism underlying dopamine cell death in Parkinson's disease. J Neurol Neurosurg Psychiat Suppl: 22-8.

Langston JW, Ballard P, Tetrud JW et al (1982) Chronic Parkinsonism in humans due to a product of meperidine analog synthesis. Science 219: 979-80

Langston JW. (1990) Predicting Parkinson's disease. Neurology 40 (Suppl): 70-4.

Mitchell JD, Jackson MJ, Pentland B. (1987) Indices of free radical activity in the cerebrospinal fluid in motor neurone disease. J Neurol Neurosurg Psychiat 50: 919-22.

Mitchell JD. (1987) Heavy metals and trace elements in amyotrophic lateral sclerosis. In: Brooks BR, ed. Motor neurone disorders and amyotrophic lateral sclerosis. Philadelphia: WB Saunders 43-60

Mitchell JD, Pentland B. (1990) Trace elements and free radicals in ALS. In: Clifford Rose F, Norris FH, eds. The etiology of ALS. Epidemiological and neurotoxicological aspects. London: Smith-Gordon & Co. 189-95.

Mitchell JD, East BW, Harris IA et al (1991) Manganese, selenium and other trace elements in the spinal cord, liver and bone in motor neurone disease. Eur Neurol 31: 7-11.

Mitchell JD. (in press) Disorders of the anterior horn cell. In: Critchley EMR, Aisen A, eds. Diseases of the spinal cord. London: Springer Verlag.

Miyata S, Nakamura S, Nagata H et al (1983) Increased manganese level in spinal cords of amyotrophic lateral sclerosis determined by radiochemical neutron activation analysis. J Neurol Sci 61: 283-93

Nagata H, Miyata S, Nakamura S et al (1985) Heavy metal concentration in blood cells in patients with amyotrophic lateral sclerosis. J Neurol Sci 67: 173-8

Pall HS, Williams AC, Blake DR et al (1987) Raised cerebrospinal fluid copper concentration in Parkinson's disease. Lancet 2: 238-41

Spencer PS, Nunn PB, Hugon J et al (1986) BMAA and BOAA: chemically related primate motor neuron toxins isolated respectively from Guamian Cycas circinalis and Indian Lathyrus sativus. Neurology 36 (Suppl 1): 135-6.

ELECTROPHYSIOLOGICAL AND FLOW CYTOMETRIC STUDIES ON MECHANISM OF GENTAMICIN NEPHROTOXICITY IN CULTURED RENAL CELLS.

H. McGlynn, N. Healy, M.P. Ryan.

Department of Pharmacology, University College Dublin, Belfield, Dublin 4. Ireland.

INTRODUCTION

The kidney is a major target organ for toxicity by many pharmaceutical compounds and environmentally important chemicals. The kidney is a complex organ with many diverse transport processes which can lead to concentration of highly toxic compounds and active metabolites. There is a need to develop better model systems in order to adapt molecular and mechanistic approaches to nephrotoxicity in order to establish site-specific processes and perturbations produced by drugs and other chemicals.

The present study concerns investigation of gentamicin nephrotoxicity in established renal cells. Aminoglycosides including gentamicin exhibit significant nephrotoxicity. Established renal cells provide useful models for study of renal pharmacology and toxicology (Gstraunthaler 1988). In this investigation we have used the established renal cell lines, LLC-PK_1 (of proximal tubular origin) and MDCK (of distal tubular and collecting duct origin). We have used biochemical, electrophysiological and flow cytometric techniques in the investigation of effects of gentamicin on these two cell types.

METHODS

LLC-PK_1 and MDCK cells were obtained from Flow Laboratories, Irvine, Scotland. Cells were maintained in Dulbeccos modified Eagles medium (DMEM) supplemented with 10% foetal calf serum, 2 mM glutamine and incubated at 37°C in an atmosphere of 95% O_2/5%CO_2. For glucose transport and viability studies, cells were grown in 24 multiwell plates. For electrophysiological and mannitol flux studies, cells were grown on millicell-HA culture plate inserts (Pitt et al. 1987). Glucose uptake was investigated using alpha-^{14}C-methyl-glucose. Transepithelial resistance of cells grown on microporous filters was measured using the millicell-HA electrical resistance system (Millipore Corp. Bedford, MA). Transepithelial flux of ^{14}C-mannitol both apical to basolateral and basolateral to apical was studied. Cell viability was assessed using double fluorescent probes propidium iodide and fluorescein diacetate in a BD FACSTAR Plus Flow Cytometer (Ormerod, 1990). Prior to flow cytometry cells were dispersed using trypsin-EDTA treatment. The effects of gentamicin at concentrations ranging from 10^{-6} - 10^{-2} M for various times of exposure ranging up to 48 hours were investigated.

RESULTS

Glucose Uptake

Glucose uptake was found to be Na-dependent and phlorizin-sensitive in LLC-PK_1 cells but not in MDCK cells. The magnitude of glucose uptake was higher in LLC-PK_1 cells with a value of 32.4 ± 4.2 nmoles/mg protein at equilibrium compared to 10.7 ± 3.4 nmoles/mg protein in MDCK. Exposure to gentamicin in the concentration range 10^{-4} to 10^{-2}M significantly reduced ($p<0.05$) glucose uptake into LLC-PK_1 cells but had little effect on glucose uptake in MDCK cells.

Transepithelial Resistances

At confluency on collagen-coated millicell-HA inserts, both LLC-PK_1 and MDCK monolayers developed transepithelial resistances. These results are shown in Table 1. In the control situation, LLC-PK_1 cells developed an electrical resistance of 143 ± 7 ohms.cm^2. This electrical resistance was very sensitive to gentamicin. At gentamicin concentrations of 10^{-6}M there was a significant reduction in transepithelial resistance to 107 ± 5 ohms.cm^2 ($p <0.05$). In LLC-PK_1 cells significant reductions in transepithelial resistances were detected at all concentrations of gentamicin. In the control situation, MDCK cells developed an electrical resistance of 252 ± 16 ohms.cm^2. The electrical resistances in these cells were less sensitive to gentamicin. Only gentamicin concentrations of 10^{-4}M or greater significantly reduced the membrane resistances in these cells.

Mannitol Fluxes

The results for apical to basolateral fluxes of mannitol are shown in Table 2. Similar results were found for the basolateral to apical fluxes of mannitol. Relatively little fluxes of mannitol occurred in the control state. However, increasing concentrations of gentamicin tended to increase the fluxes of mannitol in both cell types and these reached statistical significance at concentrations of 10^{-4}M or greater of gentamicin.

Cell Viability

The effect of gentamicin at the same concentrations and times of exposure that affected membrane resistances and mannitol fluxes were investigated on both sub-confluent and confluent cells. Gentamicin in the concentration range 10^{-6}-10^{-2}M and duration of exposure up to 48 hours did not significantly affect cell viability. Cell viability was expressed at percentage of control. Control values were assigned 100%.

Even at concentrations of gentamicin up to 10^{-2}M for 48 hours cell viability was 93 ± 10 in LLC-PK_1 and 94 ± 9 in MDCK cells.

Table 1. The effect of gentamicin on the transmembrane resistances of LLC-PK_1 and MDCK cells.

Gentamicin Conc (M)	LLC-PK_1	MDCK
Control	143 ± 7	252 ± 16
10^{-6}	107 ± 5*	267 ± 10
10^{-5}	105 ± 5*	231 ± 17
10^{-4}	101 ± 6*	176 ± 14*
10^{-3}	117 ± 11*	166 ± 16*
10-2	94 ± 4*	168 ± 14*

Results are expressed as ohms.cm^2 and given as mean ± s.e.mean; n = 14. *Indicates statistically significant difference from control p<0.05.

Table 2. The effect of gentamicin on the apical to basolateral flux of ^{14}C-mannitol in LLC-PK_1 and MDCK cells.

Gentamicin Conc (M)	LLC-PK_1	MDCK
Control	0.50 ± 0.22	0.14 ± 0.04
10^{-6}	0.65 ± 0.20	0.08 ± 0.04
10^{-5}	1.22 ± 0.15	0.22 ± 0.06
10^{-4}	0.99 ± 0.10*	1.51 ± 0.21*
10^{-3}	1.50 ± 0.60*	1.81 ± 0.07*
10-2	1.83 ± 0.50*	1.87 ± 0.02*

Results are expressed as pmoles ^{14}C-mannitol transported and are given as mean ± s.e.mean; n=6. *Indicates statistically significantly different from control p<0.05.

DISCUSSION

The results of glucose uptake experiments indicate that the LLC-PK_1 cells retain some characteristics of the proximal tubule while the MDCK cells did not exhibit Na-dependent glucose transport. Therefore, these two cell types exhibit some different transport properties reflecting their origins from different nephron sites. The sensitivity to gentamicin of glucose uptake in LLC-PK_1 cells is an interesting finding and suggests that part at least of the toxicity of gentamicin may be produced by membrane mediated events. Boogaard et al. (1989) have indicated that α-methylglucose uptake may be a sensitive parameter for mechanistic studies of acute toxicity by xenobiotics. The results with transmembrane resistances and mannitol fluxes also indicate that membrane-mediated events may play an important role in toxicity of gentamicin. Under control conditions, the resistances obtained for the

$LLC-PK_1$ and MDCK cells are compatible with those reported in the literature (Gstraunthaler, 1988). Two different strains of MDCK cells with greatly different resistances have been reported. Our results indicate that we are dealing with strain II. Gentamicin significantly reduced the transepithelial resistances in both cell types. However, the $LLC-PK_1$ cells were much more sensitive to gentamicin in this respect. The results with mannitol fluxes also support the concept that membrane-mediated events may be important in the mechanism of gentamicin toxicity. In both cell types, good monolayers were formed and there was little transepithelial flux of mannitol in the absence of gentamicin. Increased concentrations of gentamicin resulted in a disruption of the monolayer barrier and enhanced fluxes of mannitol.

The possibility existed that the changes detected may have been secondary to cytoxicity and cell death. For this reason, we examined cell viability using flow cytometry. No significant decrease in cell viability was detected at concentrations of gentamicin up to 10^{-2}M for 48 hour exposure. The results indicated that gentamicin affected membrane properties and functions including integrity of the epithelial cell layer prior to any significant effects on cell viability. These changes may in part be mediated by effects on calcium metabolism. We have previously reported that gentamicin inhibited calcium transport in brush border membrane vesicles from proximal tubular cells (Godson & Ryan 1987).

ACKNOWLEDGEMENTS

This work was supported by EOLAS, the Irish Science and Technology Agency.

REFERENCES

Boogaard PJ, Mulder GJ, Nagelkerke JF (1989) Isolated proximal tubular cells from rat kidney as in vitro model for studies on nephrotoxicity. II α-methylglucose uptake as a sensitive parameter for mechanistic studies of acute toxicity by xenobiotics. Toxicol Appl Pharmacol 101: 144-157

Godson C, Ryan, MP (1987) Effects of gentamicin on calcium transport in rat renal brush border membrane vesicles. Br J Pharmacol 91: 490P

Gstraunthaler GSA (1988) Epithelial cells in tissue culture. Renal Physiol Biochem 11: 1-42

Ormerod MG (1990) Flow cytometry: a practical approach. Oxford University Press, Oxford, P 265

Pitt AM, Gabriels JE, Badmington F, McDowell J, Gonzales L, Waugh ME (1987) Cell culture on a microscopically transparent microporous membrane. Biotechniques 5: 162-170

SELECTIVE CELLULAR TOXICITY AFTER EXPOSURE TO 4-IPOMEANOL, BROMOBENZENE (BrBz) AND 4-(METHYLNITROSAMINO)-1-(3-PYRIDYL)-1-BUTANONE (NNK) IN RAT LUNGS

R. Becher, G. Brunborg, M. Låg, J.A. Holme, E.J. Søderlund, E. Dybing and P.E. Schwarze

National Institute of Public Health, Geitmyrsveien 75, N-0462 Oslo 4, Norway

INTRODUCTION

The many different cell types identified in the lungs show a broad variation in morphology, function and metabolic capacity. Because of this variation certain cell types are more susceptible to the toxic insult of chemical exposures. The bronchiolar Clara cells and the alveolar type II cells are known to express relatively high levels of the P450 system compared to other pulmonary cell types (Boyd 1977; Philpot and Smith 1984; Minchin 1983). These cell types also appear to be target cells for several chemicals activated by P450 dependent reactions (Boyd and Reznik-Schüller 1984). Furthermore, Clara and type II cells are able to proliferate. Of particular interest for activation of a number of known lung carcinogens seems to be the expression of the P450 IIB1 gene in Clara and type II cells making them possible primary target cells for initiation of lung carcinogenesis (Belinsky et al. 1989; Becher et al. 1991).

In the present study we have examined the extent of DNA damage measured by alkaline elution in isolated rat Clara and type II cells as well as in macrophages after in vitro exposure to the activation dependent pulmonary toxicants bromobenzene and 4-ipomeanol and the pulmonary carcinogen NNK. In addition we have used light and electron microscopy to study pulmonary morphology after in vivo exposure to the same compounds.

MATERIALS AND METHODS

In vitro experiments

Clara cells and type II cells were isolated by enzymatic digestion of perfused lungs with protease followed by centrifugal elutriation. This isolation method is slightly modified from that described for rabbits (Devereux and Fouts 1981). Alveolar macrophages were collected by lavage and further purified by centrifugation. These isolated lung cells were incubated with the test chemicals for a period of 1 hr. After exposure, the cell viability was assayed by exclusion of trypan blue and found to be between 80-90%. In subsequent steps including centrifugation and resuspension in buffer, the cells were incubated with test chemicals and then prepared for analysis of DNA damage by an automated alkaline elution system (Brunborg et al. 1988). The DNA content in each of the eluted fractions was measured fluorimetrically.

For expression of the amount of DNA damage, the Normalized Area Above the Curve (NAAC) value (Brunborg et al. 1991) is presented.

In vivo experiments

In these experiments, either NNK in a dose of 10 mg/kg bodyweight or 4-ipomeanol 12 mg/kg bodyweight was injected intraperitoneally. At different time intervals the rats were sacrificed and the lungs fixed in either glutaraldehyd for electronmicroscopy or formaldehyde for light microscopy.

Due to its volatile character the exposure to bromobenzene was through inhalation of either 250 ppm or 1000 ppm for 2 hours. The rats were then sacrificed at different time intervals and the lungs prepared for either light- or electron microscopy.

RESULTS AND DISCUSSION

Our results demonstrate that both Clara and type II cells are able to metabolize NNK into DNA damaging metabolites in vitro causing single strand breaks and/or alkali labile sites, in contrast to alveolar macrophages (Fig. 1). In light of the differential expression of P450 isozymes in different lung cell types, the results indicate that the main reason for the difference in susceptibility to DNA damage between Clara and type II cells on one side and alveolar macrophages on the other is the activation by this enzyme system.

Other reports have indicated a higher expression of P450 IIB1 in Clara cells than in type II cells (Devereux et al. 1988). In contrast we found DNA damage in these two cell types to be of similar magnitude. Our results could partly reflect differences in the capacity to repair the NNK induced DNA damage in Clara and type II cells. However, we have recently found a similar expression of mRNA for P450 IIB1 in these cells (Låg et al., 1991). This could point to a similar capacity for NNK activation in Clara and type II cells.

When exposing Clara and type II cells to either 4-ipomeanol or bromobenzene, two activation dependent toxic compounds, we also demonstrated that these compounds are able to cause some DNA damage. However, for bromobenzene, the increase in DNA damage compared to the controls was only marginal. For 4-ipomeanol the increase in DNA damage was statistically significant. To our knowledge no DNA damage or genotoxic effects have previously been demonstrated after exposure to either bromobenzene or 4-ipomeanol. Moreover, apparently 4-ipomeanol caused DNA damage predominantly in isolated Clara cells and only at the highest concentration in isolated type II cells (Fig. 2). Although previous studies have shown that 4-ipomeanol can cause damage to other organs than the lungs (Kehrer and Kacew 1985), the Clara cells have been reported to be the primary cell type of the lung which is damaged (Boyd and Reznik-Schüller 1984). When comparing the alkaline elution results obtained with these two pulmonary cell types, both having a relatively high capacity to form reactive intermediates, it is interesting to note that the Clara cells are much more susceptible to DNA damage than the type II cells. This is presumably related to differential expression of P450 isozymes involved in the 4-ipomeanol metabolism in these cell types.

In the in vivo experiments no gross pathological changes were observed in the exposed animals. The light microscopic studies demonstrated no obvious pathological changes in the lungs in any of the treatment groups. By electron microscopy no obvious pathological changes were observed in the rats exposed to NNK.

However, in the rats exposed to bromobenzene, 1000 ppm, selective damage seen as vacuolisation of the cytoplasma occured in the bronchiolar Clara cells as reported previously (Dahl et al. 1990). Presumably this vacuolisation was due to a dilation of the endoplasmatic reticulum which eventually coalesced into larger entities. However, the vacuoles could also be caused by accumulation of water in the cell as a response to membrane alterations. The alveolar type I and type II cells seemed unaffected.

In comparison to the control group (Fig. 3), electron microscopic studies of the rats exposed to 4-ipomeanol indicated cytopathological lesions in the Clara cells and the neighboring ciliated cells seen as vacuolisation of the cytoplasma (Fig. 4). However, the vacuolisation observed in this exposure group seemed to be somewhat more predominant in the Clara cells, indicating that the initial cell injury after the 4-ipomeanol exposure occured in the Clara cells.

If we consider the metabolic capacity of the Clara cells and the necessity for 4-ipomeanol to be metabolically activated to become toxic, one could envision the initial formation of reactive cytotoxic intermediates in the Clara cells with a subsequent transport of reactive metabolites into the neighbouring cells, as a mechanism for damage in ciliated cells. Whether DNA damage is involved in 4-ipomeanol-induced cytotoxicity is at present not clear.

CONCLUSION

The present result from in vitro studies on DNA damage in isolated cells and in vivo morphological studies indicate that selective toxicity to Clara and type II cells are observed after exposure to either NNK, 4-ipomeanol or bromobenzene.

REFERENCES

Becher R, Låg M, Schwarze PE, Brunborg G, Søderlund EJ, Holme JA (1991) Chemically induced DNA damage in isolated rabbit lung cells (submitted for publication).

Belinsky SA, White CM, Trushin N, Hecht SS (1989) Cell specificity for the pulmonary metabolism of tobacco-specific nitrosamines in the Fisher rat. Carcinogenesis, 10, 2269-2274.

Boyd HR (1977) Evidence for the Clara cell as a site of cytochrome P450 dependent mixed-function oxidase activity in lung. Nature, 269, 713-715.

Boyd MR, Reznik-Schüller HM (1984) Metabolic basis for the pulmonary Clara cell as a target for pulmonary carcinogenesis. Toxicol. Pathol., 12.

Brunborg G, Holme JA, Søderlund EJ, Omichinski JG, Dybing E (1988) An automated alkaline elution system: DNA damage induced by 1,2-dibromo-3-chloropropane in vivo and in vitro. Analyt. Biochem., 174, 522-536.

Brunborg G, Holme JA, Søderlund EJ, Dybing E (1990) Organspecific genotoxic effects of chemicals: the use of alkaline elution to detect DNA damage in various organs of in vivo-exposed animals. Progr. Clin. Biol. Res., 340D, 43-52.

Dahl JE, Becher R, Aarstad K, Nilsen OG, Dybing E (1990) Species differences in short term toxicity from inhalation exposure to bromobenzene. Arch. Toxicol., 64, 370-376.

Devereux TR, Fouts JR (1981) Xenobiotic metabolism by alveolar type II cells isolated from rabbit lung. Biochem. Pharmacol., 30, 1231-1237.

Devereux TR, Anderson MW, Belinsky SA (1988) Factors regulating activation and DNA-alkylation by 4-(N-methyl-N-nitrosamino-1-(3-pyridyl)-1-butanone and nitrosodimethyl-amine in rat lung and isolated lung cells, and the relationship to carcinogenicity. Cancer Res., 48, 4215-4221.

Kehrer JP, Kacew S (1985) Systematically applied chemicals that damage lung tissue. Toxicology, 35, 251-293.

Låg, M, Brunborg G, Becher R, Holme JA, Søderlund EJ, Huitfeldt HS and Schwarze PE (1991) Characterization and metabolic activation of carcinogens in Clara cells and type II pneumocytes from rats. Abstract, 1991 EUROTOX Congress, Maastricht, The Netherlands.

Minchin RF, Boyd MR (1983) Localization of metabolic activation and deactivation systems in the lung: significance to the pulmonary toxicity of xenobiotics. Ann. Rev. Pharmacol. Toxicol., 23, 217-238.

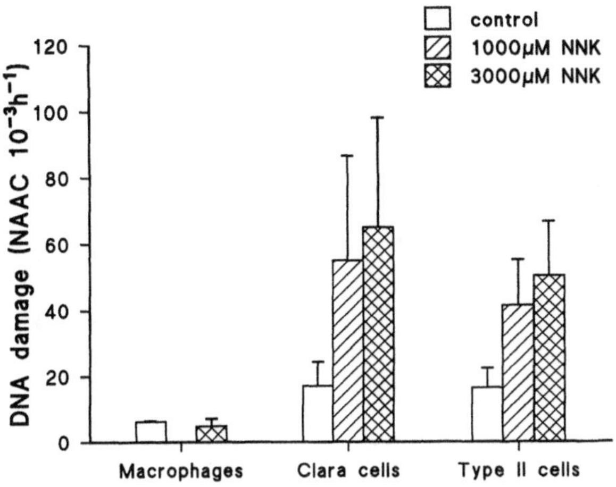

Fig. 1: DNA damage in isolated rat pulmonary cells after NNK exposure

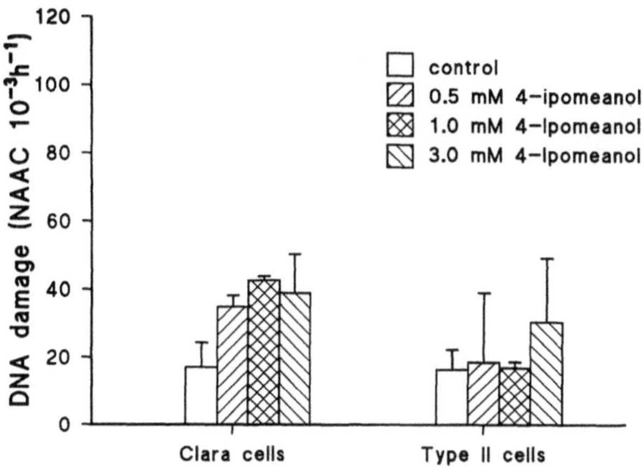

Fig. 2: DNA damage in isolated rat pulmonary cells after 4-ipomeanol exposure

Fig. 3: Control 4000x

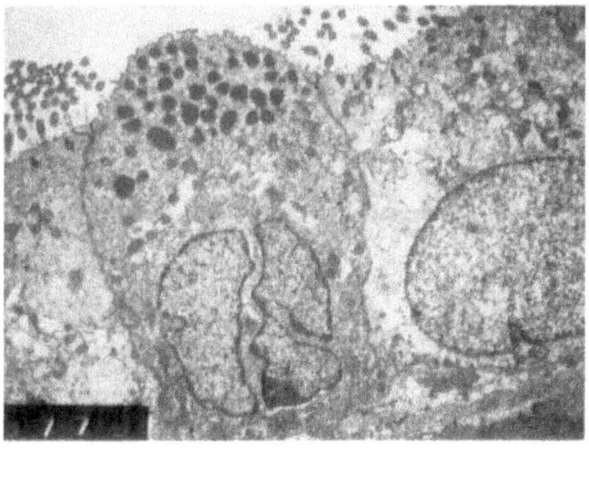

Fig. 4: 4-ipomeanol 5000x

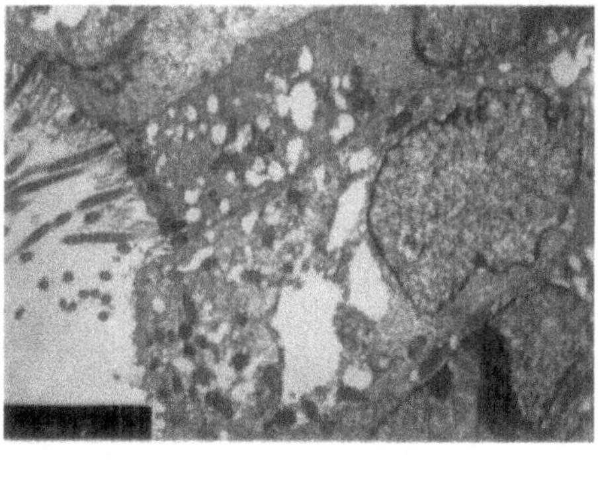

Mechanisms of Toxicity and Disease

(Chair: W. N. Aldridge, Guildford)

Decreased Hepatic Phosphoenolpyruvate Carboxykinase Gene Expression after 2,3,7,8-Tetrachlorodibenzo-*p*-dioxin Treatment: Implications for the Acute Toxicity of Chlorinated Dibenzo-*p*-dioxins in the Rat

B.U. STAHL[1,2], D.G. BEER[1,3], L.W.D. WEBER[1,2], M. LEBOFSKY[1] and K. ROZMAN[1,2]

[1]Dept. of Pharmacol., Tox. and Therap., University of Kansas Med. Ctr., Kansas City, KS 66103 (U.S.A.)
[2]GSF - Institut für Toxikologie, Section of Environmental Toxicology, D-8042 Neuherberg (F.R.G.)
[3]present address: Thoracic Surgery Research Lab., University of Michigan, Ann Arbor, MI 48109 (U.S.A.)

SUMMARY

Decreased activity of the the rate limiting gluconeogenic enzyme, phosphoenolpyruvate carboxykinase (PEPCK), has been recently suggested to be the critical lesion in the acute toxicity of 2,3,7,8-tetrachlorodibenzo-*p*-dioxin (TCDD). We now show that other toxicologically relevant chlorinated dibenzo-*p*-dioxins (CDDs), with chlorine substituents in the crucial 2-,3-,7-, and 8-positions, exert the same effect on PEPCK activity. The doses required to cause this enzyme inhibition are within the acutely toxic range for each homologue, suggesting the same mechanism of action for these compounds. To further investigate the mechanism whereby dioxins decrease PEPCK activity, Northern analysis was performed using a cDNA probe complementary to a portion of the PEPCK mRNA. We could demonstrate that after TCDD treatment hepatic PEPCK mRNA was decreased by as much as 90% compared to pair-fed control animals (day 8 after dosing). This decrease in PEPCK mRNA was paralleled by a decrease of the amount of PEPCK protein and enzymatic activity. These results indicate that the physiological changes which occur in TCDD-treated animals (decreased feed consumption, low plasma insulin and elevated plasma corticosterone levels) which under normal conditions increase PEPCK gene expression and enzyme activity, are not effective in stimulating PEPCK synthesis in TCDD-treated animals.

INTRODUCTION

The ultimate cause of death in acute TCDD-toxicity is a severe final hypoglycemia (Gorski 1987, 1988a,b). There are three sources for blood glucose, viz. dietary intake, glycogen and gluconeogenesis from amino acids or lactate. Because TCDD-treated animals gradually reduce their feed intake until they nearly cease to eat (Weber 1991b), they consequently also exhaust their glycogen and fat stores (Rozman 1986). Thus, in this late stage of intoxication their only source of energy is gluconeogenesis. We have previously shown that this pathway is impaired in TCDD-treated rats (Gorski 1990) and that this decrease is due to a dose-dependent inhibition of PEPCK activity (Weber 1991a,b). Since PEPCK is the rate limiting enzyme in the gluconeogenic pathway (Granner 1985), we suggested that this derailment of energy/glucose metabolism may be a critical lesion in acute TCDD-toxicity (Weber 1991b).

To further support this assumption and to characterize the mechanism by which TCDD causes a reduction in the enzymatic activity of PEPCK we studied the effects of three additional CDDs (with chlorination in the 2,3,7,8-positions) on PEPCK activity to establish correlations with their acute toxic potency. In addition we analyzed PEPCK activity, mRNA and the amount of PEPCK protein during a time course of 2 hours to up to 8 days after dosing. As a control for changes in PEPCK gene expression due to reduced feed intake in TCDD-treated animals, pair-fed control animals were also included in this study (8 day time point only).

MATERIALS and METHODS

The four different CDDs (2,3,7,8-tetrachlorodibenzo-*p*-dioxin; 1,2,3,7,8-pentachlorodibenzo-*p*-dioxin; 1,2,3,4,7,8-hexachlorodibenzo-*p*-dioxin; 1,2,3,4,6,7,8-heptachlorodibenzo-*p*-dioxin) used in this study were obtained from Cambridge Isotope Lab. Inc. (Woburn, MA). If not mentioned otherwise all biochemicals were purchased from Sigma Co., St. Louis, MO. Buffer salts (analytical grade) were obtained from Fischer Scientific Co., St. Louis, MO. The PEPCK cDNA fragment PC 116 (Beale 1985) and the PEPCK antiserum (goat-anti-PEPCK) for Northern and Western analyses were generously provided by Dr. D. Granner, Vanderbilt University, Nashville, TN, USA.

Animals and Treatment: Male Sprague-Dawley rats were obtained from Sasco (Omaha, NE), housed in wire bottom cages and kept on a controlled light/dark cycle (06:00 to 18:00 hr lights on). The rats had free access to tap water and rodent chow (5001, Ralston Purina, St. Louis, MO). The different CDD-homologues were dissolved in corn oil/acetone (95/5, v/v) and administered by gastric intubation. For the acute toxicity study at least 3 dose levels for each of the four homologues were used and administered to groups of 5 rats. LD20, 50 and 80 values were determined for each compound by probit analysis using the obtained mortality data. These values (LD20, 50 and 80 for each homologue) were then used as doses in a PEPCK activity dose-response study 8 days after dosing. The dose chosen in the time course study (3 rats per time point, 2, 4, 8, 16, 48, 96 and 192 hours after treatment) for Northern and Western analyses was 125 µg/kg TCDD.

Assay for PEPCK (EC 4.1.1.32) Activity: PEPCK activity was determined according to a method described by Petrescu (1979), with minor modifications as reported previously (Weber 1991a).

Isolation of mRNA and Northern Blot Hybridization: RNA was isolated by an acid guanidium thiocyanate-phenol-chloroform extraction according to Chomczynski (1987) using the RNAzol™ B Method (Cinna/Biotecx Lab. Inc. Friendswood, TX). Ten µg of cellular RNA were separated by electrophoresis in 1.2% agarose gels containing 2.2 M formaldehyde as described earlier (Beer 1988). The RNA was then transferred to nylon membranes (Gene Screen Plus, NEN Research Products, Boston, MA) with a vacuum blotting system (Pharmacia LKB 2016 Vacu Gene, Piscataway, NJ). Nick translation with deoxycytidine 5'-triphosphate [α-^{32}P] (NEN Research Products, Wilmington, DE) was used to label the PEPCK cDNA fragment. Northern prehybridization and hybridization of the PEPCK gene transcript were performed as described previously (Beer 1988). The abundance of the 28S ribosomal RNA was taken for normalization of the RNA load on the membranes and determined by direct hybridization according to the method of Barbu (1989). Membranes were exposed to XAR 5 x-ray film (Eastman Kodak Co., Rochester, NY) with intensifying screens at -80° C until an appropriately exposed autoradiogram was obtained.

Western Immunoblot Analysis of Rat Hepatic PEPCK: For analysis of the amount of PEPCK protein, frozen tissue samples were thawed in buffer (0.32 M sucrose, 10 mM K_2HPO_4 and 1mM EDTA), homogenized with a glass-teflon pestle homogenizer and 100,000 x g supernatants were prepared. Protein concentrations were determined by the method of Bradford (1976) using BSA as a standard. Samples (10 µg of protein per lane) were separated by electrophoresis in 7.5% acrylamide gels and then transferred electrophoretically (Towbin 1979) from the gel to a transfer membrane (Immobilon PVDF, Millipore Co., Bedford, MA) according to the procedure of Nielsen (1982). Immunostaining of the membranes was carried out using a primary antiserum (goat anti-rat PEPCK) at a concentration of 1:400. Following incubation with a peroxidase-labeled secondary anti-goat IgG (1:200, Cappel, Organon Technica Co., West Chester, PA), the membranes were developed with a tetramethylbenzidine (TMB) membrane peroxidase substrate kit (Kirkegaard & Perry Lab. Inc., Gaithersburg, MD) to visualize primary antibody binding. TMB-stained protein bands (Western blots) and autoradiographic signals (Northern blots) were quantified by scanning densitometry (Shimadzu CS 9000 U, Delta Instrument Co.,Waterloo, IL).

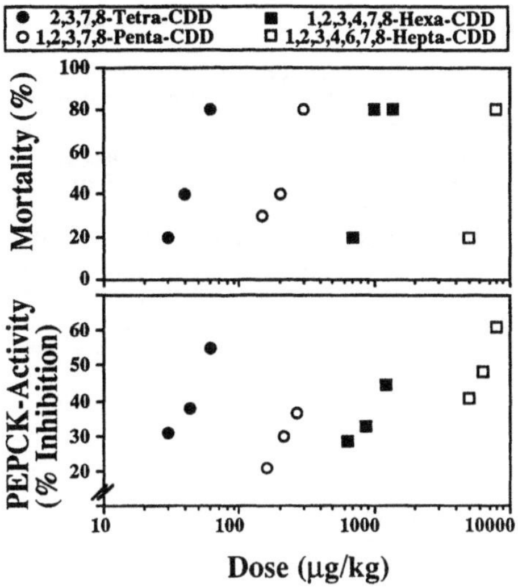

Fig 1: Dose-response relationships of mortality (upper panel) and PEPCK activity (lower panel) 8 days after treatment with 4 different CDDs.

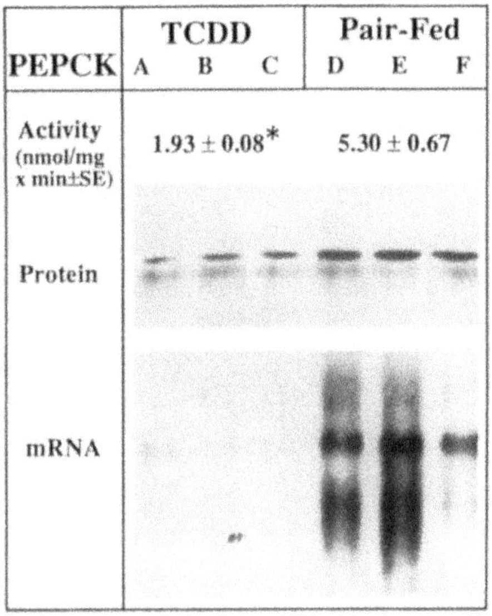

Fig 2: Effects of TCDD (125 μg/kg) 8 days after dosing on liver PEPCK activity, protein and mRNA.*significantly different from pair-fed controls ($p < 0.01$).

RESULTS AND DISCUSSION

We examined the effects of four CDDs with chlorination in the crucial 2-, 3-, 7-, and 8-positions and found that these homologues displayed the same dose-response and structure-activity relationship with regard to their acute toxic potency as well as inhibition of PEPCK activity (Fig 1). The LD50 values obtained for the different CDDs were 43, 206, 887 and 6325 μg/kg for tetra-CDD, penta-CDD, hexa-CDD and hepta-CDD, respectively. Very similar values for inhibition of PEPCK activity (38.0, 29.7, 33.7 and 48.4%) were obtained with these doses of the different congeners when assayed 8 days after dosing (Fig 1).

This inhibition of PEPCK activity was due to reduced hepatic levels of its mRNA, followed by decreased protein synthesis, beginning at approximately four hours and reaching maximum inhibition by 4 days after treatment (Fig 2, 3). In Fig 2, three TCDD-treated animals and their pair-fed controls are compared 8 days after treatment. PEPCK activity and protein at this time point were reduced by more than 60% whereas PEPCK mRNA was reduced by as much as 90%. The time course of these changes is shown in Fig 3. In this experiment ad libitum-fed animals were utilized as appropriate controls because there was no substantial reduction of feed intake at the earlier time points. Total starvation for 24 hours considerably increased the amount of PEPCK mRNA, protein and activity in control animals, but only slightly increased these parameters in 24 h starved TCDD-treated rats on day 8 (data not shown). This suggests that PEPCK gene transcription rather than PEPCK mRNA turnover is altered by TCDD.

PEPCK gene expression is under sensitive hormonal control, with insulin and glucocorticoids as key regulators (Granner 1985). Both hormones, insulin which is decreased (Gorski 1987) and corticosterone which is increased (Gorski 1988) after TCDD treatment, are altered in a manner that PEPCK gene expression should be stimulated. As shown in the present study, however, this is not the case.

On the contrary, TCDD treatment renders the organism apparently non-responsive to the physiological stimuli which, under normal conditions, stimulate PEPCK gene expression.

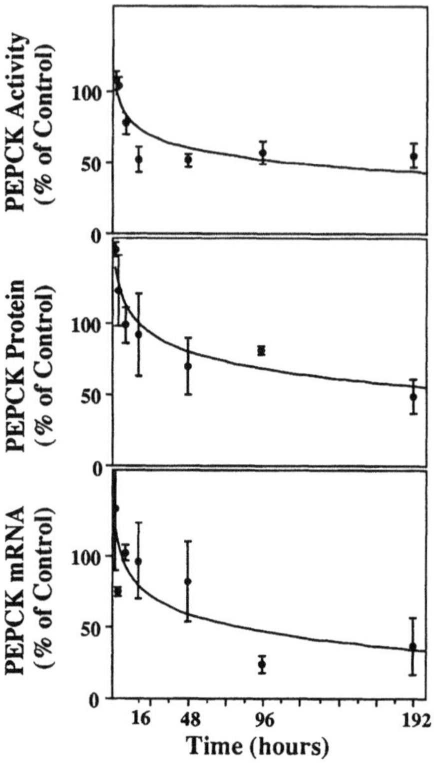

Fig 3: Time course of PEPCK activity, protein and mRNA levels in rat livers 2, 4, 8, 16, 48, 96 and 192 h after dosing compared to ad libitum-fed control animals.

Yet PEPCK enzyme activity is affected much earlier by TCDD because the hormonal changes do not occur before days 4 to 8 (Gorski 1987, 1988). This implies that these hormonal changes are secondary to enzyme inhibition and therefore are a response of the organism attempting to stimulate gluconeogenesis (Weber 1991b).

The mechanism by which TCDD alters the apparent responsiveness of the PEPCK gene to physiological stimuli is unknown. A partial explanation may be provided by the finding that other enzymes of intermediary metabolism which are regulated by glucocorticoids are also inhibited after TCDD treatment, which is possibly due to decreased ligand binding to the rat hepatic glucocorticoid receptor by as much as 40% after TCDD treatment (Sunahara 1989). The Ah receptor is not likely to be involved. All documented effects of TCDD which are proven to be cause-effect related to Ah receptor binding reach their maximum response far below acutely toxic doses, e.g. microsomal enzyme induction (Kitchin 1979), and do not correlate with different susceptibilities regarding acute toxicity in different rat strains (Pohjanvirta 1988).

So far the only biochemical effect which satisfies the criteria of occuring within the same dose range as that of acute toxicity and which also displays the appropriate structure-activity relationships for the different CDDs is inhibition of PEPCK activity.

In summary, the regulation of PEPCK activity is impaired by four different CDDs with dose-responses and structure-activity relationships almost identical to acute toxicity. The reduced enzymatic activity of PEPCK is apparently due to reduced transcription of PEPCK mRNA leading to reduced protein synthesis. Furthermore, the regulation of the PEPCK-gene is altered in a fashion which causes dioxin-treated animals to be non-responsive to endogenous signals which under normal conditions induce PEPCK gene expression.

ACKNOWLEDGMENT

This work was supported by the Bundesministerium für Forschung und Technologie of the Federal Republic of Germany. Dr. Stahl also gratefully acknowledges the training support provided by Deutsche Forschungsgemeinschaft (Sta 300/2-1).

REFERENCES

Barbu V, Dautry F (1989) Northern blot normalization with a 28S rRNA oligonucleotide probe. Nucleic Acids Res 17: 115

Beale EG, Chrapkiewicz NB, Scoble HA, Metz RJ, Quick DP, Noble RL, Donelson JE, Biemann K, Granner DK (1985) Rat hepatic cytosolic phosphoenolpyruvate carboxykinase (GTP). J Biol Chem 260: 10748-10760
Beer DG, Neveu MJ, Paul DL, Rapp UR, Pitot HC (1988) Expression of c-*raf* protooncogene, gamma-glutamyltranspeptidase, and gap junction protein in rat liver neoplasms. Cancer Res 48: 1610-1617
Bradford MM (1976) A rapid and sensitive method for the quantitation of microgram quantities of protein using the principle of protein dye binding. Anal Biochem 72: 248-254
Chomczynski P, Sacchi N (1987) Single-step method of RNA isolation by acid guanidium thiocyanate-phenol-chloroform extraction. Anal Biochem 162: 156-159
Gorski JR, Rozman K (1987) Dose-response and time course of hypthyroxinemia and hypoinsulinemia and characterization of insulin hypersensitivity in 2,3,7,8-tetrachlorodibenzo-p-dioxin (TCDD)-treated rats. Toxicology 44: 297-307
Gorski JR, Muzi G, Weber LWD, Pereira DW, Arceo RJ, Iatropoulos MJ, Rozman K (1988a) Some endocrine and morphological aspects of the acute toxicity of 2,3,7,8-tetrachlorodibenzo-p-dioxin (TCDD). Toxicol Pathol 16: 313-320
Gorski JR, Muzi G, Weber LWD, Pereira DW, Iatropoulos MJ, Rozman K (1988b) Elevated plasma corticosterone levels and histopathology of the adrenals and thymuses in 2,3,7,8-tetrachlorodibenzo-p-dioxin treated rats. Toxicology 53: 19-32
Gorski JR, Weber LWD, Rozman K (1990) Reduced gluconeogenesis in 2,3,7,8-tetrachlorodibenzo-p-dioxin (TCDD)-treated rats. Arch Toxicol 64: 66-71
Granner DK, Andreone TL (1985) Insulin modulation of gene expression. Diabetes/Metabolism Rev 1: 139-170
Kitchin KT, Woods JS (1979) 2,3,7,8-tetrachlorodibenzo-p-dioxin (TCDD) effects on hepatic microsomal cytochrome P-448-mediated enzyme activities. Toxicol Appl Pharmacol 47: 537-546
Nielsen PJ, Manchester KL, Towbin H, Gordon J, Thomas G (1982) The phosphorylation of ribosomal protein S6 in rat tissues following cycloheximide injection, in diabetes, and after denervation of diaphragm. J Biol Chem 257: 12316-12321
Petrescu I, Bojan O, Saied M, Barzu O, Schmidt F, Kühnle HF (1979) Determination of phosphoenolpyruvate carboxykinase activity with deoxyguanosine 5'-diphosphate as nucleotide substrate. Anal Biochem 96: 279-281
Pohjanvirta R, Juvonen R, Kärenlampi S, Raunio H, Tuomisto J (1988) Hepatic Ah-receptor levels and the effect of 2,3,7,8-tetrachlorodibenzo-p-dioxin (TCDD) on hepatic microsomal monooxygenase activities in a TCDD-susceptible and -resistant rat strain. Toxicol Appl Pharmacol 92: 131-140
Rozman K (1989) A critical view of the mechanism(s) of toxicity of 2,3,7,8-tetrachlorodibenzo-p-dioxin. Dermatosen 37(3): 81-92
Rozman K, Pereira D, Iatropoulos MJ (1986) Histopathology of interscapular brown adipose tissue, thyroid, and pancreas in 2,3,7,8-tetrachlorodibenzo-p-dioxin (TCDD)-treated rats. Toxicol Appl Pharmacol 82: 551-559
Sunahara GI, Lucier GW, McCoy Z, Bresnick EH, Sanchez ER, Nelson KG (1989) Characterization of 2,3,7,8-tetrachlorodibenzo-p-dioxin-mediated decreases in dexamethasone binding to the rat hepatic cytosolic glucocorticoid receptor. Mol Pharm 36: 239-247
Towbin H, Staehelin T, Gordon J (1979) Electrophoretic transfer of proteins from polyacrylamide gels to nitrocellulose sheets: procedure and some applications. Proc. Natl. Acad. Sci. USA 76: 4350-4354
Weber LWD, Lebofsky M, Greim H, Rozman K (1991a) Key enzymes of gluconeogenesis are dose-dependently reduced in 2,3,7,8-tetrachlorodibenzo-p-dioxin (TCDD)-treated rats. Arch Toxicol 65: 119-123
Weber LWD, Lebofsky M, Stahl BU, Gorski JR, Muzi G, Rozman K (1991b) Reduced activities of key enzymes of gluconeogenesis as possible cause of acute toxicity of 2,3,7,8-tetrachlorodibenzo-p-dioxin (TCDD) in rats. Toxicology 66: 133-144

CO-CULTURES OF RAT HEPATOCYTES IN THE STUDY OF VALPROATE TOXICITY.

Rogiers V, Akrawi M*, Vandenberghe Y, Shephard E*, Phillips I**, Vercruysse A

Department of Toxicology, Vrije Universiteit Brussel, Brussels, Belgium
* Department of Biochemistry and Molecular Biology, University College London, London, UK
** Department of Biochemistry, Queen Mary and Westfield College, London, UK.

INTRODUCTION

The use of sodium valproate (VP), an effective anti-epileptic drug, has been associated with irreversible hepatotoxicity of which the mechanism has not yet been elucidated (Eadie 1988, Cotariu 1988).

This short paper deals with one specific aspect of liver toxicity, namely the eventual inducing capacities of VP on xenobiotic biotransformation systems of cultured rat hepatocytes and of their co-cultures with rat epithelial cells. Activities of key enzymes of xenobiotic phase I and II biotransformation and cytochrome P-450 subfamily gene expression have been studied using both in vitro models.

METHODS

- Hepatocytes were isolated from 24 h fasted male Wistar rats (200-300g) and brought in culture (Rogiers 1986) and co-culture (Guguen-Guillouzo 1986) as previously described. For both culture systems three conditions were studied : controls, cells treated daily with 25, 50, 100 and 200 µg/ml medium and cells treated daily with 2mM phenobarbital (PB) as positive controls.
- Cytochrome P-450 content was determined according to Omura (1964) with the modifications introduced by Paine (1979).
- 7-Ethoxycoumarin O-deethylase (ECOD) (Rogiers 1986), aldrin epoxidase (AE) (Rogiers 1988) and glutathione S-transferase (GST) (Habig 1974) were measured.
- Proteins were determined with a Bio-Rad protein assay kit.
- mRNA analysis was carried out using Northern blot hybridisation and RNase protection assays with probes specific for P450IIB1, P450IIB2 mRNA's. These methods have been described in detail elsewhere (Akrawi 1991).
- The results were statistically analyzed using repeated measurement analysis of variance with 2 variables (treatment, culture time) (Norusis 1986).

RESULTS AND DISCUSSION

In figs 1, 2 and 3 the effects of different VP concentrations on the cytochrome P-450 content, ECOD- and GST-activities, respectively, of co-cultured rat hepatocytes are shown.

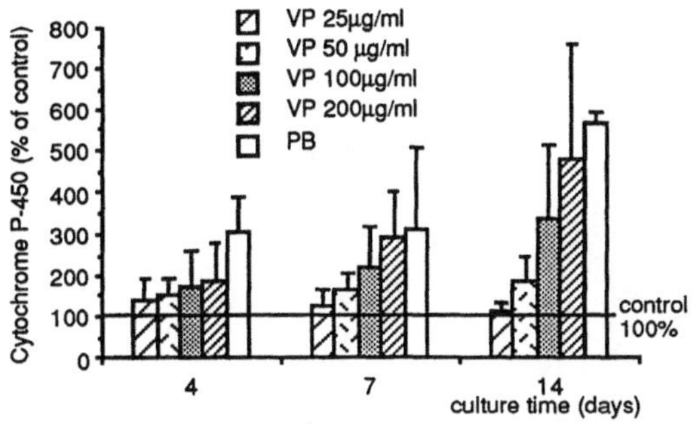

Fig. 1 : Cytochrome P-450 content of co-cultured rat hepatocytes treated with different VP concentrations and with 2mM PB.
For each time interval controls are taken as 100% values and the results for the treated cells are expressed against these (n=4).

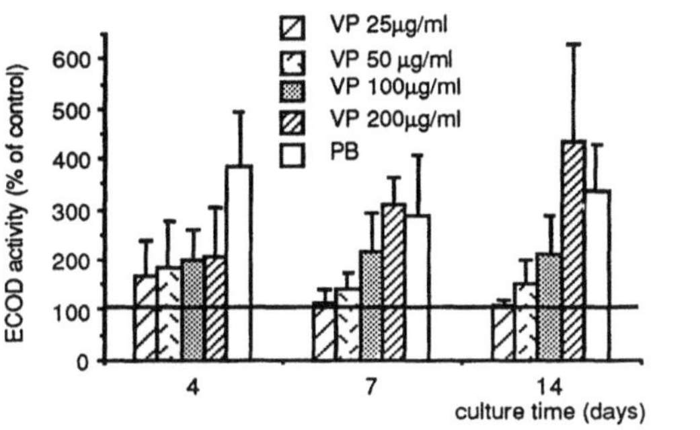

Fig. 2 : ECOD activity of co-cultured rat hepatocytes treated with VP and PB. The results of treated cells are expressed as percentages of the controls, taken as 100% values (n=4)

It is clear that the cytochrome P-450 content and the ECOD activity of VP treated hepatocytes increase as a function of VP concentration ($P<0.01$). On the contrary, GST activity is significantly decreased after VP treatment ($P<0.05$) but an effect of the concentration can also be shown ($P<0.001$). The same conclusion can be reached for pure cultures of hepatocytes, treated with VP (results not shown), although the results are statistically less significant probably because of the shorter exposure time to VP. In both in vitro models, the expression of AE activity is, under control conditions, rapidly and nearly completely lost making it impossible to study the effects of VP on AE activity (Rogiers 1990). In fig 4 Northern blot hybridisation is shown for the P450IIB subfamily in a 7 days old co-culture.

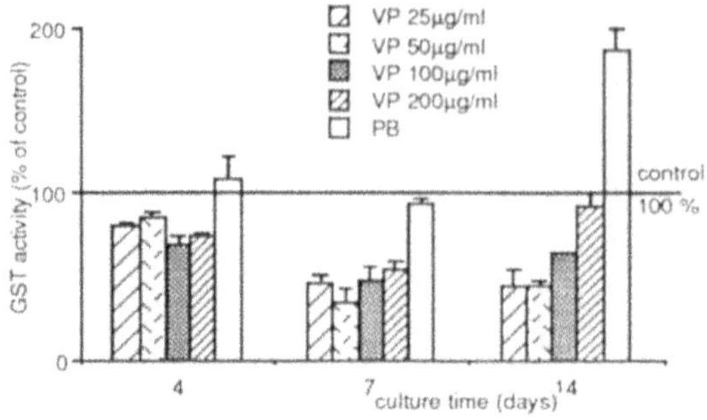

Fig. 3 : GST activity of co-cultured rat hepatocytes treated with VP and PB. The results of treated cells are expressed as percentages of the controls, taken as 100% values (n=3).

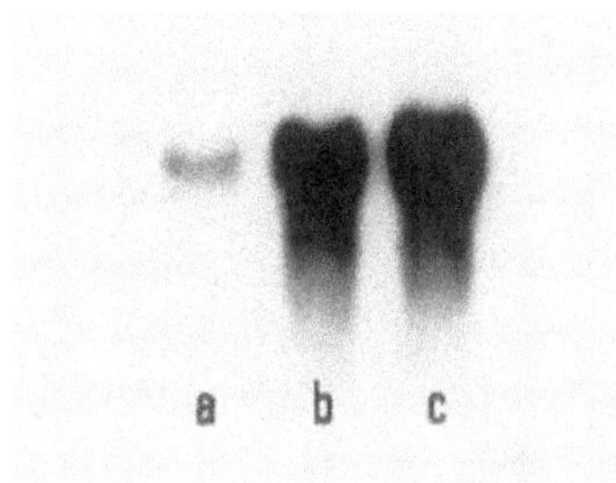

Fig. 4 : Northern blot hybridisation analysis of P450IIB mRNA.
Total RNA was extracted either from hepatocytes co-cultured for 7 days in the absence of drugs (a), in the presence of 2 mM PB (b) or VP (200 µg/ml) (c).
12 µg RNA was loaded on each track.

In contrast to pure cultures (results not shown here), P450IIB mRNA is present in co-cultures for the whole culture period. It is inducible by PB for at least 2 weeks with the maximum being obtained after 7 days. Also, VP significantly increases the amount of P450IIB mRNA by about 6-fold (fig 4, track c) which is comparable to the induction obtained by PB (fig 4, track b). RNase protection assays, using specific antisense RNA probes, distinguishing between P450IIB1 and P450IIB2 mRNA's, show that both, P450IIB1 and P450IIB2 mRNA's are induced by VP in a very comparable way to the effect obtained after PB treatment (results not shown here).

In conclusion, from previous in vivo results (Rogiers 1988) and these in vitro data in co-cultured hepatocytes, it appears that VP has, in rat liver, a complex effect on xenobiotic biotransformation systems including the induction of certain cyt P-450 isoenzymes such as P450IIB1 and P450IIB2.

REFERENCES

Akrawi M, Rogiers V, Vandenberghe Y, Palmer C, Vercruysse A, Shephard E, Phillips I (1991) Maintenance and induction in co-cultured hepatocytes of components of the cytochrome P-450 mediated mono-oxygenase. Biochemical Journal submitted.

Cotariu D, Zaidman JL (1988) Valproic acid and the liver. Clinical Chemistry 34 : 890-897

Eadie MJ, Hooper WD, Dickinson RG (1988) Valproate associated hepatotoxicity and its biochemical mechanisms. Medical Toxicology 3 : 85-106

Guguen-Guillouzo C, Clement B, Baffet G, Beaumont C, Morel-Chany E, Glaise D, Guillouzo A (1983) Maintenance and reversibility of active albumin secretion by adult rat hepatocytes co-cultured with another liver epithelial cell type. Experimental Cell Research 143 : 47-54

Habig WG, Pabst MJ, Jakoby WB (1974) Glutathione S-transferases. The first enzymic step in mercapturic acid formation. The Journal of Biological Chemistry 249 : 7130-7139

Norusis MJ (1986) Storing Memories, Repeated Measures, Analysis of Variance. In : SPSS-PC+, Advanced Statistics. SPSS Inc., Chicago USA, p B152-B181.

Omura T, Sato R (1964) The carbon monoxide - binding pigment of liver microsomes. The journal of Biological Chemistry 239 : 2370-2378

Paine AJ, Lesley JW, Legg RF (1979) Apparent maintenance of cytochrome P-450 by nicotinamide in primary cultures of rat hepatocytes. Life Sciences 24 : 2185-2192

Rogiers V, Adriaenssens L, Vandenberghe Y, Gepts E, Callaerts A, Vercruysse A (1986) Critical evaluation of 7-ethoxycoumarin O-deethylase activity measurements in intact isolated rat hepatocytes. Xenobiotica 16 : 817-828

Rogiers V, Vandenberghe Y, Callaerts A, Sonck W, Maes V, Vercruysse A (1988) The inducing and inhibiting effects of sodium valproate in vivo on the biotransformation systems of xenobiotics in isolated rat hepatocytes. Xenobiotica 18 : 665-673

Rogiers V, Vandenberghe Y, Callaerts A, Verleye G, Cornet M, Mertens K, Sonck W, Vercruysse A (1990) Phase I and phase II xenobiotic biotransformation in cultures and co-cultures of adult rat hepatocytes. Biochemical Pharmacology 40 : 1701-1706

SELENIUM-SILVER INTERACTION IN MICE

V.Eybl, M.Koutenská, J.Koutenský, J.Sýkora, R.Bláhová, V.Smolíková
Department of Pharmacology, Charles University Faculty of Medicine, CS-301 66 Pilsen, ČSFR

INTRODUCTION

It has already been shown that the essential trace element selenium has great potential as a factor in the maintenance of human health. It may also change some biological functions e.g. by means of interaction with various other essential elements. Selenium is effective in the prevention of toxic effects resulting from exposure to cadmium, mercury, lead and other metals (Högberg and Alexander 1986; Nève 1991).

In our previous experiments the absolute antagonism between sodium selenite and cadmium or mercury was demonstrated. The whole body burden of these metals was increased and the tissue distribution of both metals changed by the administration of selenium (Eybl et al 1975). On the other hand the retention of selenium was increased mainly due the administration of mercury (Eybl et al 1969). The kinetics of selenium may be directly influenced by the interaction of this element with zinc (Eybl et al 1986).

Also the mutual antagonism between silver and selenium has been shown by many investigators. However, only a few papers deal with the effect of silver on selenium metabolism and kinetics (Högberg and Alexander 1986; Fowler and Nordberg 1986). The distribution of selenium in the body after the simultaneous i.v. administration of sodium selenite and silver nitrate was studied only in one acute short term experiment (Naganuma et al 1983).
In the present study the effect of silver on the acute toxicity of selenium, as well as the influence of silver on the whole body retention and tissue distribution of selenium at different time intervals, was examined.

MATERIALS AND METHODS

The experiments were performed in male mice (ICR, 22-25g b.w.) divided into groups of animals (see Tables).
The following chemicals were used:
$Na_2{}^{75}SeO_3$ (with carrier), Rotop-Germany), silver acetate a.g. (Lachema, ČSFR), sodium selenite.5 H_2O a.g. (Merck, Germany). The radioactivity in the whole body and in various organs after the i.v. administration of $Na_2{}^{75}SeO_3$ (0.58 mg/kg) was measured by using a scintillation spectrometer with NaJ (Tl) detector. The survival of the mice was recorded on the 10th day of the experiment. All results were statistically evaluated. The values given are means and SEM.

RESULTS AND DISCUSSION

The protective effect of a nontoxic dose of silver acetate in cases of acute intoxication with sodium selenite was demonstrated (Table 1).

Table 1. Effect of simultaneous administration of silver acetate on survival of mice receiving sodium selenite[1]

Group	10 day survival ratio	% survival	significance
Na selenite	0/12	0	-
Silver acetate only	12/12	100	-
Na selenite with Ag acetate	8/16	50	$p < 0.01$

[1]Ag acetate was administered i.p. (20mg/kg) at Ag:Se ratio 1:1 immediately after s.c. administration of Na_2SeO_3

Pretreatment with a single dose of silver acetate also increased the whole body burden of selenium (Table 2).The content of selenium in most of the organs continued to be raised on the 14th day of the experiment. The highest accumulation of selenium occurred in the liver, gastrointestinal tract, spleen and in the testes. The rate of selenium elimination was markedly reduced due to silver administration. The effectiveness of silver acetate is of the same order as that of mercury.The effect of silver on selenium kinetics and toxicity is probably based on the formation of inert silver selenide. Our results, including those published previously, correspond to the model for metal selenide formation under biological conditions proposed by Nuttal (1987). Aaseth et al (1969) analyzed silver deposits in a case of iatrogenic argyria. These deposits in the kidney consisted of Ag_2Se.

CONCLUSIONS

The protective effect of silver acetate in acute intoxication with sodium selenite was demonstrated. The pretreatment with a single dose of silver acetate increased the whole body burden of selenium. The level of selenium in various organs was raised. This effect is still visible many days after the administration of the compounds and is probably due to the formation of inert silver selenide.

REFERENCES:

Aaseth J, Olsen A, Halse J, Hoving T (1981) Argyria - tissue deposition of silver as selenide. Scand J clin Lab Invest 4: 247-251

Fowler BA, Nordberg GF (1986) Silver. In: Friberg L, Nordberg GF, Vouk V (eds) Handbook on toxicology of metals, 2nd edn. Elsevier Scientific Publishers BV, p 521

Högberg J, Alexander J (1986) Selenium. In: Friberg L, Nordberg GF, Vouk V (eds) Handbook on toxicology of metals, 2nd edn. Elsevier Scientific Publishers BV, p 482

Eybl V, Sýkora J, Koutenská M, Koutenský J, Mertl F (1969) The

Table 2. Whole body burden and tissue content of selenium in mice at various time intervals after the administration of Na_2SeO_3/Na_2SeO_3 with Ag-acetate[1,2]

Time interval:	24 h		7th day		14th day	
Ag-acetate:	0	+	0	+	0	+
Whole body burden:	24.34(0.88)	44.09(2.14)[4]	16.66(0.83)	34.19(2.02)[4]	10.54(0.37)	26.12(1.54)[4]
Liver:	5.65(0.17)	11.07(0.35)[4]	5.32(0.29)	10.93(0.28)[4]	2.68(0.07)	8.84(0.11)[4]
Kidneys:	2.06(0.26)	2.73(0.16)	1.69(0.17)	1.23(0.14)	0.58(0.02)	1.36(0.05)[4]
Brain:	0.07(0.01)	0.07(0.01)	0.07(0.01)	0.06(0.02)	0.04(0.02)	0.11(0.01)[3]
Gastroint.tract:	2.88(0.16)	5.90(0.32)[4]	1.59(0.14)	3.61(0.25)[4]	1.03(0.08)	3.17(0.22)[4]
Spleen:	0.18(0.02)	0.32(0.01)[4]	0.09(0.01)	0.20(0.02)[4]	0.05(0.01)	0.20(0.02)[4]
Heart:	0.11(0.01)	0.17(0.02)[3]	0.08(0.01)	0.11(0.02)	0.03(0.01)	0.07(0.03)
Lungs:	0.52(0.11)	0.66(0.04)	0.30(0.04)	0.31(0.02)	0.15(0.02)	0.41(0.05)[3]
Testes:	0.15(0.01)	0.27(0.03)[3]	0.27(0.04)	0.46(0.04)[4]	0.22(0.02)	0.71(0.09)[4]
Femurs:	0.12(0.03)	0.31(0.01)[4]	0.10(0.01)	0.14(0.01)[3]	0.04(0.01)	0.01(0.01)[3]

[1] Ag-acetate was administered i.p. at Ag:Se ratio 1:1 immediately after i.v. administration of $Na_2{}^{75}SeO_3$ (0.58 mg/kg i.p.). The mice were divided into 2 groups of 15 animals (control = Ag-acetate: 0; experiment = Ag-acetate: +). Always 5 mice of each group were sacrified at the following time intervals: 24 h, 7 and 14 days. The whole body burden was always measured in all living animals.

[2] The results are expressed as % of dose applied; values represent the mean and SEM (in parentheses)

[3] $p < 0.05$

[4] $p < 0.01$

influence of zinc, cadmium and mercury on the retention and distribution of selenium in mice. Physiol bohemosl 18: 336-337
Eybl V, Sýkora J, Koutenský J, Koutenská M, Mertl F (1975) In vivo interaction of mercury and cadmium with sodium selenite. Proc Eur Soc Toxicol 16: 262-269
Eybl V, Sýkora J, Mertl F (1986) In vivo interaction of selenium with zinc. Acta pharmacol toxicol 59: S7, 547-548
Naganuma A, Tanaka T, Maeda K, Matsuda R, Tabata-Hanyu J, Imura N (1983) The interaction of selenium with various metals in vitro and in vivo. Toxicology 29: 77-86
Nève J (1991) Physiological and nutritional importance of selenium. Experientia 47: 187-193
Nuttal KL (1987) A model for metal selenide formation under biological conditions. Medical Hypotheses 24: 217-221

Environmental Impact

(Chair: W. SEINEN, Utrecht)

ENVIRONMENTAL IMPACT ON DOMESTIC AND WILD ANIMALS : FIELD DATA FROM A VETERINARY ANTI-POISON CENTER

Gérard KECK
National Center of Veterinary Toxicological Informations
Veterinary School of Lyon
69280 Marcy l'Etoile - France

The National Center of Veterinary Toxicological Informations is the only Veterinary Poison Center in Europe, open on a 24 hour-basis to veterinarians, but also to all persons concerned with problems of animal toxicology in the field, for both domestic and wild animals.

It is based on the activity of a team a veterinary teachers, post-graduate and pre-graduate students and is in close relation with a laboratory of toxicological analyses. Since 1981, the data have been stocked and treated with a computerized data bank (Keck et al., 1982 ; Lechenet et al., 1986)

From the experience of our Center, field data concerning the exposure and impact on animals may help for a better knowledge of the hazards for the animals populations themselves, but also for detecting and monitoring the risks for human health and for the environment.

From this view, animals play an important role of "sentinels". Animals are often more directly and permanently exposed to the pollution sources, by inhalation, or eating contaminated forage, or drinking contaminated water, etc.....

This **"veterinary toxico-vigilance"** may concern complex pollutions (dumping sites, incinerators, industrial effluents, ...) as well as more precise chemical agents such as pesticides, PCBs, heavy metals, etc ...

Acute or chronic cases of intoxication reported in animals concretize the major types of risks, including those where contaminated animals may represent, through animal productions, a vector of chemical risks for man.

1) Acute risks of animal intoxications

- The day to day work of our Center is to bring assistance for acute intoxications (or suspicions). Cattle and dogs are the most affected species because of their number but also of their avidity and propension to swallow almost every thing. The major class of chemicals involved is by far pesticides, which represent more than 50 % of the calls (Lorgue et al., 1985).
It should be stressed that an important proportion of those cases is due to **improper and uncontrolled uses of agro-chemicals** including very often intentional, criminal applications against domestic animals (especially dogs and cats) or wild species (chemical poaching of fish or of protected mammals and prey-birds).

Strychnine is still used for that criminal purpose, but **carbamate insecticides** (aldicarbe, methiocarbe, carbofuran) are more and more frequently found.
Packaging or stocks of **obsolete agrochemicals** may also lead to acute intoxications, often severe because of the usually high toxicity of those old products ; several cases of arsenicals poisonings were reported in cows, after the farmer had removed old stocks from his barn and dropped it in a corner of a pasture.

- **Acute mortalities of fishes,** for which our Center is often called, are often due to accidental or negligent spills in rivers or lakes. One of the most frequent case is the spill of wood preservatives, containing usually lindane and pentachlorophenol from the wood-industries, which are surpringly unaware of the toxicity of these chemicals.
At the higher level of **chemical disasters,** massive mortalities of fish and water birds did reveal the River Rhine pollution of Schweitzer-halle in Basel, following the fire of the Sandoz depot of pesticides (including organo-mercurials, organo-phosphates, nitrophenols) and its extinction by the firemen which carried the chemicals into the river.
In this major accident, the action of a "veterinary fire man" (a french speciality) was quite efficient by making observations (the swans showed a pink colour on their feather), adapted samples of fish and taking rapid conter-measures for protection of animal and public health in collaboration with other health professions.

* **Fires of waste disposals** may represent an acute risk for animals pasturing around the site. The combustion of banal wastes may release toxic substances, such as chlorine (from PVC recipients) or cyanhydric acid (from the polymers of car seats).
We were consulted in July 1990 for a collective and brutal mortality of young cattle, 6 days after such a waste fire, with severe gastro-enteritis and dehydration. The veterinary authorities and of course, the farmer, were convinced this was due to a toxic release from the fire. In fact, numerous autopsies, analyses, clinical and epidemiological considerations showed the the episode was due to a brutal parasitic invasion due to Ostertagia larvae.
This shows that animals may be "false sentinels" of pollution, due to confusion with many acute diseases. Moreover, possibilities of indemnisation are important bias, for acute as well as for chronic accidents, which will be seen in the next part.

2) Chronic accidents in animals

The most problematic chemicals are indeed **bio-cumulative agents** (which is by far not the case of all chemicals).

2.1. Heavy metals

The case of Mercury brings a strong illustration of the risks for human health through the food-chains, animals being vectors of chemical risks (Keck, 1979).
The Minamata accident was due to the accumulation of methyl-mercury in the fish and shell-fish, which induced severe neuropathies in man

and in newborns. One of the first sign of the dramatic episode was the neurologic changes observed in cats (ataxy, perpetual crying) which ate only fish and are generally very susceptible to neuro-toxic impact of chemicals.

Industrial pollution from smelters or metallurgical factories is often detected by the chronic impact on domestic animals pasturing contaminated grass near the factory.
Lead is a major source of intoxications in cattle : industrial pollution was well studied in the North of France, from atmospheric origin or from disposal of highly contaminated mud or smelter-sand. But on a general scale painting is by far the most common source of intoxication in cattle (Keck, 1987).
In water-fowl, the ingestion of lead-shots at the bottom of the ponds represent a real threat of saturnism, as was shown by several studies in France and in many other countries.

Mixed industrial pollution with zinc, copper and probably cadmium induced arthropathies in horses in the region of Normandie, since 1985, of the same type as described by Gunson et al. (1982).

2.2. Organic derivatives

Undoubtely, the "model-pollutants" are **chlorinated hydro-carbons**, especially **PCBs** and **chlorodioxins**, which present a very different toxicological status (Monot and Keck, 1982).
PCBs show high persistance and bio-accumulation but a moderate toxicity. We investigated several cases of high levels of PCB contamination, especially in cow milk, which is an interesting witness of basic or accidental contamination by organic pollutants.
In 1988, the milk of a dairy farm showed levels higher as 50 ppm (lipid weight), due to spills of a small plant which treated old electric condensators. The fish in the river 1 km downstream had 10 to 20 ppm.
However, no clinical or zootechnic impact was found on the animals; nor did the farmer's family which consumed several months their own milk show any clinical or bio-chemical changes (but the investigations were quite limited, due to psychological considerations).

Another recent episode of pollution by PCBs concerned the River Rhône upstream from Lyon ; high levels, up to 5 ppm, were found in the fishes, associated with a significant increase of cytochrome P 450-enzymatic activities in the liver (Monot et al., 1990). Fishing in this area was forbidden. The origin of the contamination was probably a plant of incineration of PCB, located several kilometers from the river.

Enzymatic activities such as cytochrome P 450 are more and more used as "biomarkers" for evaluating exposure and impact of contamination by pollutants such as PAHs (polycyclic aromatic hydrocarbons) or chlorinated hydrocarbons, which are effective enzyme inducers, in wild or domestic animals.

Chlorodioxins are certainely pollutants of high concern. They are often associated with chlorophenols and may be generated through the combustion of chlorinated wastes, especially during waste incineration.
During the Seveso accident (1976), italian veterinarians reported many cases of affected cattle and mortalities in domestic rabbits ; their incidence correlated quite well with the level of exposure (Mantovani Al., 1989).

Several episodes of **chlorodioxins or chlorofuranes intoxications** were reported in animals, often associated with exposure to chlorinated hydrocarbons such as chlorophenols or PCBs. Chickens showed edemas and hepato-renal degeneration, after picking the plastic coverage of metallic fence.
In Missouri, a collective intoxication of horses with dermal and hepatic lesions was due to wood-sawing of the horseman-ship sprayed with recuperation-oil (Buck).
Humans were also exposed, an epidemiological study conducted between 1972 and 1982 showed an increase in fetal and peri-natal mortality in the exposed population, but the difference was not statistically significant by comparing with a control population (Stockbauer, 1988).

The **impact of chlorinated hydrocarbons on reproduction** is probably a sensitive marker of the long-term risks of an industrial pollution, for example near an incinerator.
Cows have been selected and followed up on reproduction criteria. So parameters such as fecondity, fertility or gemellity may be studied in exposed versus non-exposed zones.
In the Bonnybridge area (Scotland), the number of twins in human was slightly but significantly increased, from 16 to 20 ‰ in the districts exposed to the incinerator towards 3 to 13 ‰ in the non-exposed districts. In the high exposure district, the level of gemellity observed in dairy cows pasturing close to the incinerator increased from 0-2 % in 1972-1979 to 8-22 % in 1980-1985. The increase of gemellity could be due to the oestrogenic effects of chlorinated hydrocarbons, such as PCBs or chlorodioxins (Lloyd et al. 1988).

Conclusion

There is undoubtely a need to increase the knowledge and information concerning the exposure and effects of domestic and wild animals to environmental impact.

During the last ten-years, efforts have been made in some countries to develop that field of activity. This should lead to the creation of veterinary toxicological centers at the national level and a good cooperation at the international level, in close relation with medical toxicological organizations.

Recently, several meetings stressed the role of veterinary activities in natural or technological disasters, which are part of the field of veterinary public health and should be included in the disaster prepared ness plans (Mantovani, Keck and Héritier, 1991).

Morever, epidemiological field research should include an animal approach : this is the case for example, for an important french program on the risks of industrial wastes on human and animal health and on the environment.

References

Buck WB (1979) Animals as monitors of environmental quality. Vet. Hum. Toxicol., 21, 4 ; 277-284

Gunson DE, Kowalczyk DF, Shoop CR (1982) Environmental zinc and cadmium pollution associated with generalized osteochondrosis, osteoporosis and nephrocalcinosis in horses. JAVMA, 180 : 295-299

Keck G, Lorgue G, Eynard H (1982) Une banque de données informatisées en toxicologie vétérinaire créée par le CNITV. Journal de Toxicologie Médicale, 2 : 93-104

Keck G (1987) Paediatric aspects in veterinary toxicology : the example of lead intoxication in the calf. Vet. Human Toxicol., 29 (suppl. 2) 25-27

Keck G, Héritier P (1991) Actions vétérinaires en situation de catastrophe Veterinary actions in disaster situations. San Marino, 16-18 november 1989. Veterinary Public Health Reports - ISS/WHO/FAO - CC/IZST e/91.13

Léchenet J, Nahas K, Keck G, Lorgue G (1986) Mise en place d'une banque de données. Une expérience vétérinaire. Le Centre National d'Informations Toxicologiques Vétérinaires. Rec. Med. Vet., 162 -6-7) ; 873-877

Lloyd O, Lloyd M, Williams FLR, Lawson A (1988) Twinning in human populations and in cattle exposed to air pollution from incinerators. Brit. J. Ind. Med., 45, 556-560

Lorgue G, Rampaud M, Keck G and Jaussaud P (1985) Toxicological data on wild and domestic animals poisoned with rodenticides. Acta Zool. Fennica, 173 : 213

Mantovani Al (1989) The role of veterinary sciences in chemical disasters and emergencies. Pre-hospital and Disaster Medicine, 4, 2 : 155-159

Mantovani A, Keck G, Cautin JF, Trenti F, Bussi B (1991) Veterinary action in disasters. CEMEC Monographs n° 5

Monod G, Keck G (1982) PCBs in Lake Geneva (Lake Leman) fish. Bull. Environ. Contam. Toxicol., 29 ; 570-576

Stockbauer JW, Hoffmann RE, Schramm WF, Edmonds LD (1988) Reproductive outcomes of mothers with potential exposure to 2, 3, 7, 8 TCDD. Am. J. Epidemiol., 128, 410-419

BLOOD MERCURY AND SERUM SELENIUM CONCENTRATIONS IN REINDEER HERDERS IN THE ARCTIC AREA OF NORTHERN FINLAND

Luoma PV, Näyhä S, Pyy L,* Korpela H, and Hassi J*

*Regional Institute of Occupational Health, and
Department of Public Health Science, University of Oulu,
SF-90220 Oulu, Finland

Environmental factors have a great impact on life in the arctic. The arctic area of northern Finland (Lapland) is surrounded by Sweden in the west, Norway in the north, and the Soviet Union in the east. In Lapland there are approximately 200000 inhabitants, including some 4000 reindeer herders who have a life style of their own. As part of a health survey we studied a group of reindeer herders for blood organic, inorganic and total mercury and serum selenium concentrations. Mercury, a heavy metal, may accumulate in the food chain and be harmful to health (Birke et al. 1972; Study Group of Minimata Disease 1968). Selenium, an essential trace element, is a constituent of glutathione peroxidase that protects cellular structures against oxidative damage (Rotruck et al. 1973).

SUBJECTS

1648 male reindeer herders, a representative sample of all reindeer herding males in Finland, were invited to a medical examination, and 163 healthy males, a random sample of the 1100 participants, were selected for this study. 101 subjects were Finns, 24 Lapps, members of the ethnic minority in the area, the rest being of both Finnish and Lapp origin. The age of participants was 41.5 ± 1.0 years (mean$\pm$SE) and body mass index 25.9 ± 0.3 kg/m^2. 26 per cent of them were smokers.

METHODS

The participants attended the local health centre for a medical examination. The blood samples were taken after an overnight fast.

Blood organic and inorganic mercury were determined by atomic absorption spectrometry as described by Lajunen et al. (1985). Serum selenium was determined by atomic absorption spectrometry equipped with a graphite furnace (Voellkopf and Grobinski 1984).

The analysis of variance and linear regression analysis were employed for the statistical treatment the data.

Table 1. Blood mercury and serum selenium levels in the northwestern (NW) (Muonio, Kolari, Kittilä) and northeastern (NE) (Utsjoki, Inari, Savukoski) communes of the arctic area of northern Finland

	Mercury nmol/l			
	Organic	Inorganic	Total	Selenium µmol/l
NW communes (n=36)				
Mean	14.3	5.1	19.4	1.42
SE	1.4	0.6	1.5	0.05
NE communes (n=82)				
Mean	27.9[a]	6.0	33.9[a]	1.50
SE	2.3	0.4	2.5	0.04

[a]$p<0.0005$

RESULTS

Blood organic, inorganic and total mercury concentrations (Mean$\pm$SE) of the whole group were 24.1 ± 1.4, 5.4 ± 0.3 and 29.5 ± 1.5 nmol/l, respectively. Nineteen subjects had blood total mercury levels exceeding the upper normal limit (50 nmol/l). Blood total mercury was closely related to organic mercury ($r=0.98$, $p<0.001$), and it increased with age ($r=0.27$, $p<0.001$). Mercury showed no association with body mass index or smoking.

Table 2. Blood mercury and serum selenium levels in the Finns and Lapps in the arctic area of northern Finland

	Mercury nmol/l			
	Organic	Inorganic	Total	Selenium µmol/l
Finns (n=101)				
Mean	19.6	5.3	24.9	1.45
SE	1.2	0.4	1.3	0.03
Lapps (n=24)				
Mean	35.2[a]	5.5	41.7[a]	1.64[b]
SE	6.4	0.9	6.9	0.05

[a]$p<0.0005$, [b]$p<0.005$

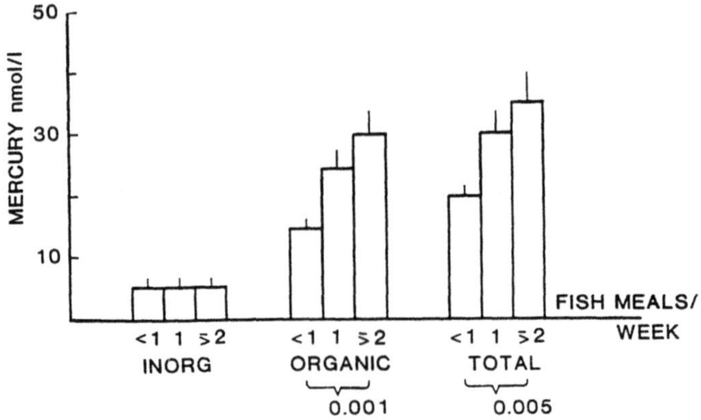

Fig. 1. Blood inorganic, organic and total mercury concentrations (mean with SE) in reindeer herders according to the number of fish meals per week in wintertime. 59 subjects had a fish meal at least twice a week and 35 subjects less frequently than once a week.

Blood organic and total mercury levels varied by living area, being higher in the northeastern communes than the north-western ones (Table 1). Table 2 shows that the Lapps had higher organic and total mercury levels than the Finns. Blood organic and total mercury concentrations increased with increasing fish consumption (Fig. 1). The association remained significant when adjusted for age.

Serum selenium (mean±SE) of the reindeer herders was 1.50± 0.02 µmol/l, ranging from 1.10 to 3.90 µmol/l. The Lapps had higher serum selenium than the Finns (Table 2).

DISCUSSION

More than ten per cent of the healthy reindeer herders had a high blood mercury concentration. The high values were more common in the northeast than northwest, and higher in the Lapps than Finns. These trends probably reflect the effect of environmental factors on the soil and water, and the dietary habits. The buffer capacity of the soil in the arctic is low. The pollution in the northeastern part of the country, close to the Kola Peninsula, probably exceeds the buffer capacity of the soil, and causes acidification. This promotes the formation of toxic compounds such as organic mercury (methylmercury) and their increase in the food chain. Dietary factors, e. g. fish consumption can hence affect mercury levels in blood and tissues.

Serum selenium varied within the normal range. High selenium concentrations were associated with high mercury levels. This probably reflects their common pathway in the food chain (Hansen 1984). Selenium is an antioxidant and may reduce the toxicity of mercury, and increase the tolerance to it.

In this study high blood mercury levels were observed in healthy subjects. Whether mercury has any association with the diseases of the reindeer herders remains to be investigated.

SUMMARY

163 healthy reindeer herders living in the arctic area of northern Finland were investigated for blood mercury and serum selenium concentrations. More than ten per cent of the subjects had blood mercury above the upper normal limit. High organic and total organic mercury levels were most common in the northeastern part of the country. High organic and total mercury and selenium values were more common in the Lapps than the Finns. Blood mercury levels increased with increasing fish consumption. The pollution in the northeastern part of this country probably leads to the increased formation of organic mercury, and its enrichment in the food chain. The association of high selenium with high mercury probably reduces the toxicity of the latter.

REFERENCES

Birke G, Johnels AG, Plantin L-O, Sjöstrand B, Skerfving S, Westermarck T (1972) Studies on humans exposed to methyl mercury through fish consumption. Arch Environ Health 25:77-91

Hansen JC, Kromann N, Wulf HC, Alboge K (1984) Selenium and its interrelation with mercury in wholeblood and hair in an East Greenlandic population. Sci Total Environ 38:33-40

Lajunen L, Kinnunen A, Yrjänheikki E (1985) Determination of mercury in blood and fish samples by cold-vapor atomic absorption and direct current plasma emission spectrometry. Atomic Spectrosc 6:49-52

Rotruck JT, Pope AL, Ganther HE, Swanson AB, Hafeman DG, Hoekstra WG (1973) Selenium: Biochemical role as a component of glutathione peroxidase. Science 179:588-590

Study Group of Minimata Disease (1968) Minimata disease. Kumamoto University, Kumamoto Japan

Voellkopf U, Grobenski Z (1984) Interference in the analysis of biological samples using the stabilized temperature platform furnace and Zeeman background correction. Atomic Spectrosc 5: 115-122

Toxipathology: Natural and Induced Syndromes in Laboratory Animals

(Chair: R. A. Ettlin, Basel;
A. Rodriguez-Consuegra, Sevilla)

Natural Syndromes in Dogs

A. Pospischil

Institute of Veterinary Pathology, University of Zurich, Winterthurer Str. 268, CH-8057 Zurich, Switzerland

Syndrome is a word widely used in medical- and veterinary medical terminology, although its definition is rather difficult.

DEFINITION

As a more general definition a syndrome consist of a number of symptoms resulting from a single cause or so commonly occurring together which collectively indicate an often abnormal condition of the body or mind (Blood and Studdert, 1988). Syndrome recognition or pattern recognition is one of the important clinical diagnostic steps during a more preliminary stage in the making of a diagnosis. This recognition is based on the comparison of the subject case and previous cases in the clinicians memory and the one is recognized as the replica of the other (Blood and Studdert, 1988).In *general pathology* the term syndrome is characterised by a combination of morphological lesions occurring simultaneously without a known etiology.

USE OF THE TERM SYNDROME IN THE LITERATURE

Medical literature data bankslist the term syndrome 54,334 times between 1986 and 1990. These references include about 30,000 different syndromes known to date. The same search period in veterinary medical data banks lists the term 2,536 times, 73 citations refer to canine syndromes.

The canine syndromes most frequently reported are:

Paraneoplastic syndromes (Morrison, 1979; Brown, 1981; Gorman, 1990; Hohlfeld, 1990) are any disorders of organ function produced by the noninvasive actions of a tumor. They appear to result from unphysiological effects occurring in distant target organs caused by the production and release into the circulation of unusual amounts of macromolecules. They may briefly be characterised as systemic effects of tumors, like hypercalcemia, caused e.g. by apocrine gland adenocarcinoma of the anal sac, or hyperhistaminemia due to mast-cell tumors or hypertension as a reaction to circulatory active hormones produced e.g. by pheochromocytomas. Some of the better known paraneoplastic syndromes are:

Cushing's syndrome, including *Morbus Cushing* (Siegel et al., 1967, O`Neill, 1979; Reznik et al.,1987) a canine hyperadrenocorticism (Fig. 1 A), in rare cases caused by corticotropin-secreting small basophilic adenomas of the pars distalis of the pituitary gland. The great majority of cases is, however, initiated by corticotrophic (chromophobic) adenomas of the pars distalis and pars intermedia (Fig. 1 B).

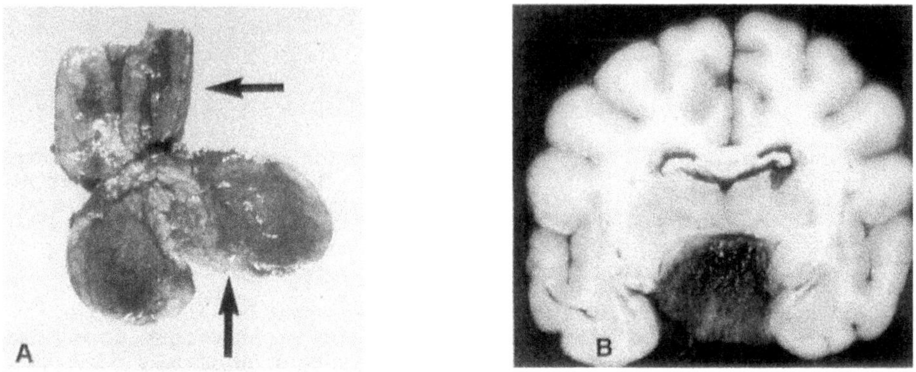

Figure 1 Adrenocortical hyperplasia (A, arrows) and adenoma of the pituitary gland (B) in a 10 year old dachshund with Cushing`s syndrome.

Zollinger-Ellison syndrome (Straus et al., 1977; Hoppe et al., 1980; English et al.,1988) in the dog as in man is characterized by peptic ulcers due to gastrin secreting pancreatic islet cell tumors or those deriving from G cells in the gastric or duodenal mucosa (gastrinomas). Metastases to the liver or hepatic lymph nodes are frequent. Peptic ulcers and reflux esophagitis in such cases result from acid hypersecretion. Gastrin together with histamine and acetylcholine stimulates the gastric acid secretion.

Other syndromes frequently reported are:

The Caval syndrome (Mason and Tait, 1969; Jackson et al., 1977; Eaton and Rosol, 1989; Kitagawa et al., 1990), is an acute obstruction of the Vena cavae by large masses of adult heartworms (Dirofilaria immitis) leading to hepatic congestion and centrolobular necrosis. On physical examination peripheral veins are engorged, the jugular veins are enlarged and shows a pulse, the heartbeat is forceful. In dogs with caval syndrome, heartworms locate from the pulmonary arteries to the Venae cavae, but most parasites are present in the right atrium and ventricle. On echocardiography, heartworm echoe moves back and forth between the right atrium and right ventricle and the function of the tricuspid valve was severely hindered. There is a consistent correlation between the severity of the clinical disease, the extent of centrolobular hepatic congestion and necrosis and the numbers of heartworms involved. This form of heartworm disease is entirely different from the usual majority of cases.

Key Gaskell syndrome in dogs (canine dysautonomia) (Pollin and Sullivan, 1986) is a degenerative disease of neurons of the sympathetic and parasympathetic ganglia. Degenerative neurons are swollen or shrunken and contain excentric or pycnotic nuclei. Difficulties in swallowing food, infrequent defecation and urinary incontinance were reported.

Flat puppy syndrome (Harkness and McCormick, 1981; Blunden, 1986; Mehl, 1990) sometimes used as a synonym to swimming puppy- or even fading puppy syndrome is characterized by a dorsoventral compression of the chest and polyarthrodysplasia (Fig. 2). They are unable to walk after day 21 of age. One of the etiological factors discussed is a insufficient floor condition in the breeding units.

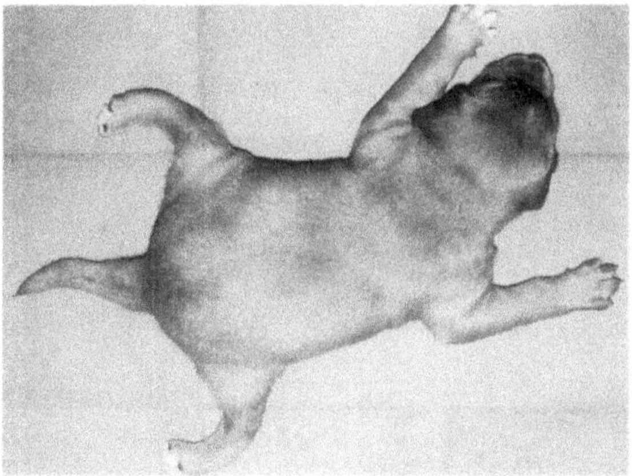

Figure 2 Ten day old Labrador puppy with typical signs of flat puppy syndrome.

Canine gastrointestinal syndrome.(Prescott et al., 1978; Burrows, 1977) is clinically characterized by a sudden onset of vomiting and bloody diarrhea. On histopathological examination the intestinal mucosa is necrotic and usually a layer of large Gram positive bacteria closely attached to the surface is visible (Fig. 3). Large numbers of Clostridium welchii were recovered in several cases.

The occurrence of these syndromes in the literature was compared with the post-mortem investigations of dogs from 1987 to 1990 at the Institute of Veterinary Pathology in Zurich. Anamnestic data could not be included since they were not regularly available.

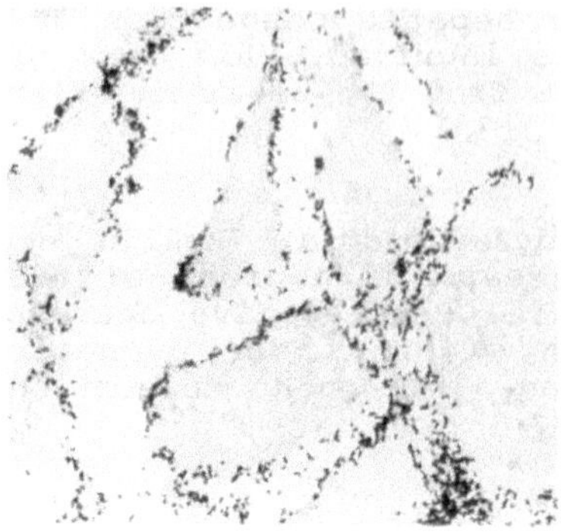

Figure 3 Histopathology of the small intestine of a dog with hemorrhagic gastrointestinal syndrome.Necrotic villi are lined by bacteria.

DATA FROM CANINE ROUTINE POST-MORTEM INVESTIGATIONS

Since 1987 records of post-mortem investigations at the Institute of Veterinary Pathology, Zurich are processed and kept as computer files (Wegmann 1991).Diagnoses are coded and stored by IBM software called Eram on Wang hardware. Since 1989 biopsy records are treated likewise, before this all records were recorded in writing.

From 1987 until 1990, 1636 canine post-mortems were performed (Fig. 4). The number per year remained relatively stable. The major disease groups are shown in Fig. 5. Tumors of different organs and varying grades of malignancy represent the largest group, followed by infectious diseases, of which the etiology could be confirmed in 17.1% of the cases.

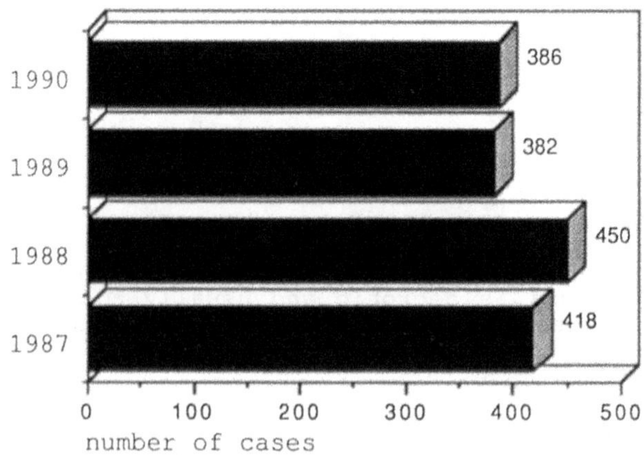

Figure 4 Number of canine post-mortem investigations 1987 - 1990

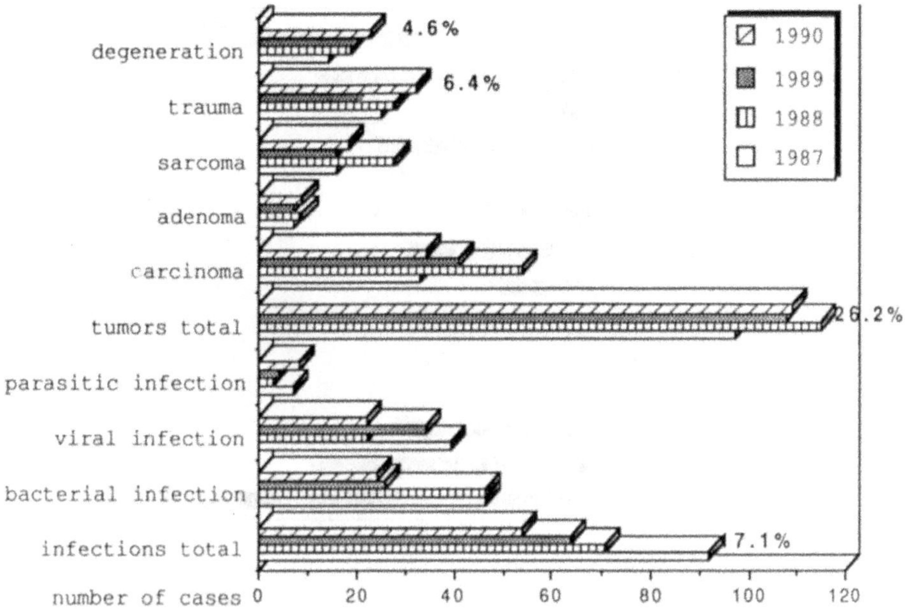

Figure 5 Major disease groups in canine post-mortem investigations 1987-1990

The number of naturally occuring syndromes in the post-mortem material presented is shown in Fig. 6. There were 24 genuine natural canine syndromes (1.22% of the total) another 15 cases were suspects. Taken together 2.4% of all post-mortems may be classified as belonging to a syndrome. Most of these are in the group of paraneoplastic syndromes which is not surprising since tumors represent 26.2% of the total diagnoses. More than half of all tumors were malignant (carcinomas and sarcomas = 14.6%).

Altogether natural syndromes represent only a small number of cases in a routine diagnostic veterinary pathology laboratory. This compares well with the 73 canine syndromes of the total 2,536 citations referenced in the veterinary literature data banks.

In a representative evaluation of 9,112 canine post-mortem investigations from 1951 to 1965 von Sandersleben et al. (1973) do not use the term. Some of the disease complexes mentioned, however, occur under different names.

The evaluation of the biopsy material is biased regarding the occurrence of natural syndromes by the fact that in many cases the anamnestic data transmitted by referring veterinarians was not complete.This means that a diagnosis is based on the material received but further conclusions may not be drawn.

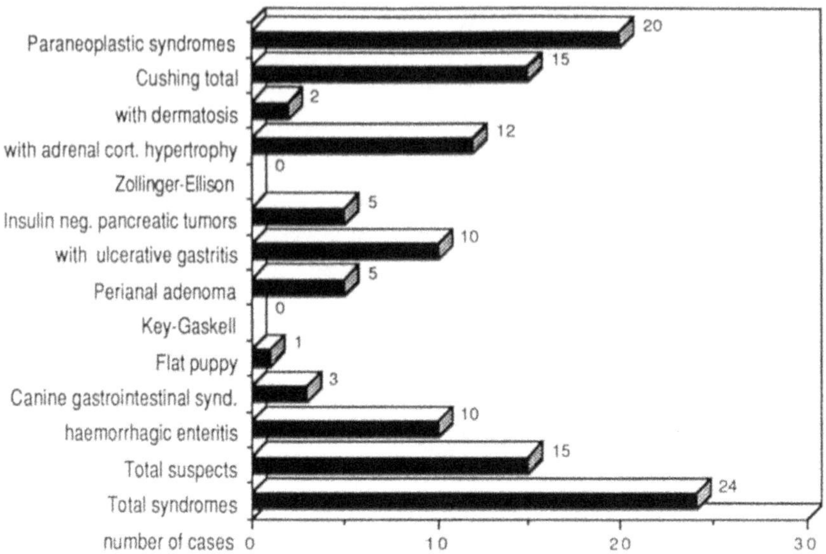

Figure 6 Natural canine syndromes in post-mortem material at the Institute of Veterinary Pathology, Zurich 1987 - 1990

From 1987 until 1990 5884 canine biopsies were seen at the Institute of Veterinary Pathology in Zurich (Fig. 3).From about half of those from 1989 and 1990 detailed diagnoses were available as computer data (Fig. 4, 5). From a total of 1496 neoplastic lesions, 434 (29%) were malignant (carcinoma or sarcoma). Most of these neoplasms probably caused paraneoplastic syndromes. Insufficient anamnestic data, however, do not allow any further conclusions.

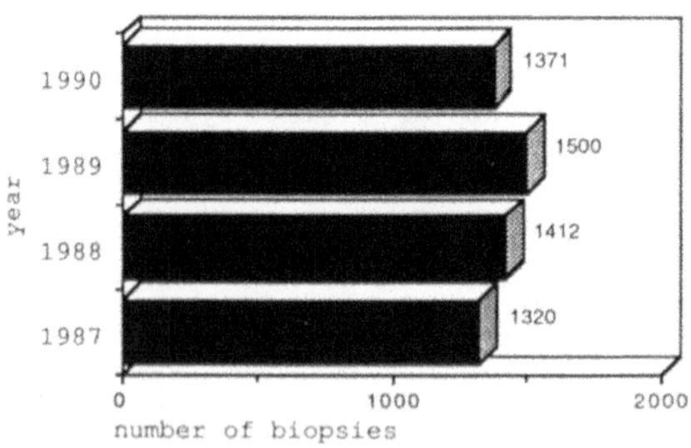

Figure 7 Number of canine biopsies investigated from 1987 - 1990

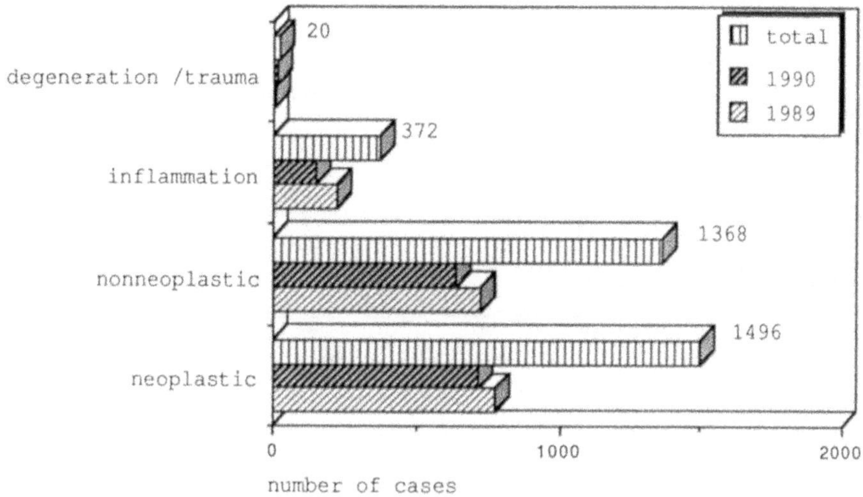

Figure 8 Major disease groups diagnosed in canine biopsies 1989-1990

REFERENCES

Blood DC, Studdert VP (1988) Baillières comprehensive veterinary dictionary. Baillière Tindall, London, p 888

Blunden AS (1986) A review of the fading puppy syndrome. Vet Ann 26: 264-269

Brown NO (1981) Paraneoplastic syndromes of humans, dogs and cats. J Amer Anim Hosp Assoc 17:911-916

Burrows CF (1977) Canine hemorrhagic gastroenteritis. J Amer Anim Hosp Assoc 13:451-458

Eaton KA, Rosol TJ (1989) Caval syndrome in a Dirofilaria immitis-infected dog treated with dichlorvos. J Am Vet Med Assoc 195: 223-224

English RV, Breitschwerdt EB, Grindem CB, Thrall DE, Gainsburg LA (1988) Zollinger-Ellison syndrome and myelofibrosis in a dog. J Am Vet Med Assoc 192: 1430-1434

Gorman NT (1990) Clinical management of tumours in geriatric dogs and cats: Systemic effects of tumours and paraneoplastic syndromes. Vet Rec 126: 395-398

Harkness JE, McCormick LF (1981) Swimming puppy syndrome in a litter of German Shepherd pups. Vet Small Anim Clin 76: 817-821

Hohlfeld R (1990) Myasthenia gravis and thymoma: Paraneoplastic failure of neuromuscular transmission. Lab Invest 62: 241-247

Hoppe RP, van der Gaag I, Lamers CBHW (1980) Zollinger-Ellison syndrome in 3 dogs. Vet Pathol 17: 177-186

Jackson RF, Seymour WG, Growney PJ, Otto GF (1977) Surgical treatment of the caval syndrome of canine heartworm disease. J Am Vet Med Assoc 171: 1065-1067

Kitagawa H, Sasaki Y, Ishihara K, Kuwahara Y (1990) Development of artificial model of caval syndrome in canine heartworm disease. Jpn J Vet Sci 52: 1029-1035

Mason RW, Tait AR (1969) Posterior caval syndrome in a dog. Aust Vet J 45:435-436

Mehl NB (1990) Fading puppy and fading kitten syndrome. Dansk Vet Tidskr 73: 750-756

Morrison WB (1979) Paraneoplastic syndromes of the dog. J Am Vet Med Assoc 175:559-561

O`Neill JA (1979) Cushing-like syndrome in a dog. Vet Med Small Anim Clin 74: 1261-1262

Pollin M, Sullivan M (1986) Canine dysautonomia resembling the Key-Gaskell syndrome. Vet Rec 118: 402-403

Prescott JF, Johnson JA, Patterson JM (1978) Haemorrhagic gastroenteritis in the dog associated with Clostridium welchii. Vet Rec 103: 116-117

Reznik M, Melon J, Lambricht M, Kaschten B, Beckers A (1987) Neuroendocrine tumor of the nasal cavity (esthesioneuroblastoma). Report on one case with paraneoplastic Cushing like syndrome. Ann Path 7: 137-139

Siegel ET, Kelly DF, Berg P (1967) Cushing`s syndrome in the dog. J Amer Vet Med Assoc 150: 760-766

Straus E, Johnson GF, Yalow RS (1977) Canine Zollinger-Ellison syndrome. Gastroent 72: 380-381

von Sandersleben J, Schäffer E, Weisse J (1973) Erkrankungs- und Todesursachen des alternden Hundes aus der Sicht der Sektionsstatistik. Kleintierpraxis 18: 25-32

Wegmann P (1991) Computer application in diagnostic veterinary pathology. Schweiz Arch Tierhlk 133:93

Inducible thyroid lesions in the rat with special reference to mediation or interactions with other organ systems.

C. Gopinath

Huntingdon Research Centre; Huntingdon; United Kingdom

Several unrelated compounds are known to induce thyroid follicular lesions in the rat, some acting directly on the epithelium and others acting indirectly. The frequent observation of hyperplastic and neoplastic changes of the thyroid follicles in chronic toxicity studies using rats is further evidence of the sensitivity of this epithelium in this species. TSH responses in man, to changes in thyroxine levels are often insignificant (Hill, Erdreich, Paynter, Roberts, Rosenthal and Wilkinson 1989). In the rats these responses are known to be prompt and strong. The increased sensitivity of rat thyroid is attributable to several species-specific features. They include; comparatively low half-life of active thyroxine; absence of thyroxine binding globulin in rat plasma to bind the free thyroxine, considerably greater plasma levels of TSH; markedly greater TSH responses and greater biliary excretion of thyroxine in rat compared to man.

Morphological lesions:

Follicular cell hypertrophy, (observed as increased height of epithelium) and hyperplasia accompanied by reduced colloid content and sometimes increased vascularity are commonly seen as induced changes in short-term toxicity studies using rats. Confirmation of a proliferative effect is often difficult by routine microscopy. Morphometric analysis is often found useful in these circumstances. The useful parameters to measure include: size and volume of thyroid gland, height and volume of follicular cells; number of follicular cells, size and number of follicles, and extent of vascularity.

With more prolonged administration of test compounds the hyperplastic responses becomes more apparent by the crowding of follicular epithelium and prominent infolding sometimes resulting in papillary ingrowths into the follicular lumina. Other changes encountered include atrophy of follicles, pigmentation of epithelium, apoptosis and changes in the appearance of colloid. Rare instances of thyroiditis are also reported. In long term (chronic) studies, cystic follicles, focal hyperplasia and cystic follicular hyperplasia are also encountered. Neoplastic changes of follicles include adenomas and adenocarcinomas.

A clear knowledge of the age related changes are important to assess induced thyroid lesions as morphological and functional parameters vary with age (Rao, Heywood and Gopinath 1991).

Classification of compounds acting on thyroid follicles

Hill et al (1989) has given a detailed classification of compounds which act on follicular epithelium.

Table 1 briefly summarises these classifications.

From this table it is clear that it is important to establish whether the compounds in question act directly on follicular epithelium by inhibiting the synthesis of follicular hormones or they act peripherally by altering the metabolism and or elimination of follicular hormones. In almost all instances in rats the follicular epithelial reactions are influenced by changes in TSH. Mediation by the liver plays an important role in peripheral metabolism of thyroid hormones. It is important to establish the mechanisms by which the compound reacts.

A series of function tests can be used to investigate these in rodent studies and some of the more routinely used tests are summarised in Table 2:

Tests 4, 5, 6 and 7 give information on the metabolism and elimiination of thyroid hormones. The tests 1-7 give a useful assessment of the functional status giving insight to the mechanistic aspects of the test compound employed. The study of the hepatocyte metabolism is essential as the hepatic microsomes possess glucuronyl transferases which mediate in the conjugation of T4 and also monoiodinase which assists in peripheral conversion of T4 into T3. The extent of hepatocellular binding and uptake of thyroxine, aspects of competetive inhibitions, and rate of biliary flow are among other important factors worthy of studying as all of these influence follicular changes in the rat.

Thyroid interactions with other organ systems

1. PITUITARY GLAND

The role of thyroid-pituitary axis is well understood and the inter-relationship is strong in all species, although species specific increased sensitivity are also well known. The surgical removal of thyroid gland in rats leads to hyperplasia and even neoplasia of TSH cells in the anterior pituitary (Kirschbaum 1957). These type of tumours are hormone dependant conditional neoplasms which alter their character on changing the hormonal status.

TSH induces thyroid growths in vivo and in vitro. Most of the induced follicular lesions in rats are mediated through TSH stimulation. The resultant hyperplasia and neoplastic changes are TSH dependant. Withdrawl of stimulation results in regression of these effects under experimental conditions (Jemec 1980).

FIGURE 1
ORGAN WEIGHTS AFTER 4 WEEKS

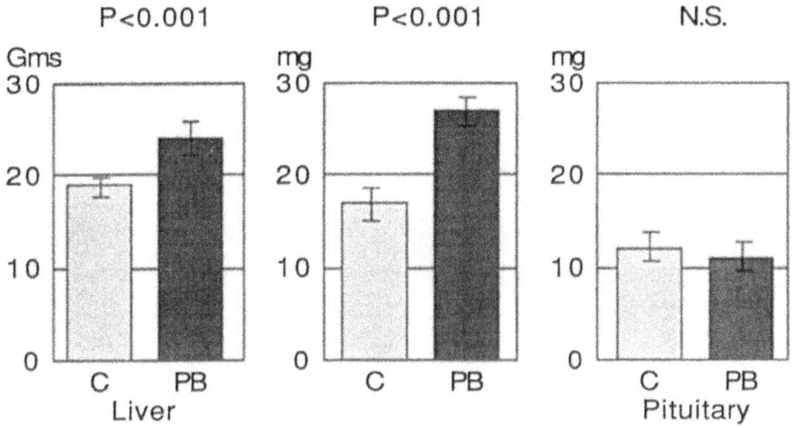

C = Control group PB = Phenobarbital treated group

FIGURE 2
HEPATIC MICROSOMAL PROTEIN
CYTOCHROME P450

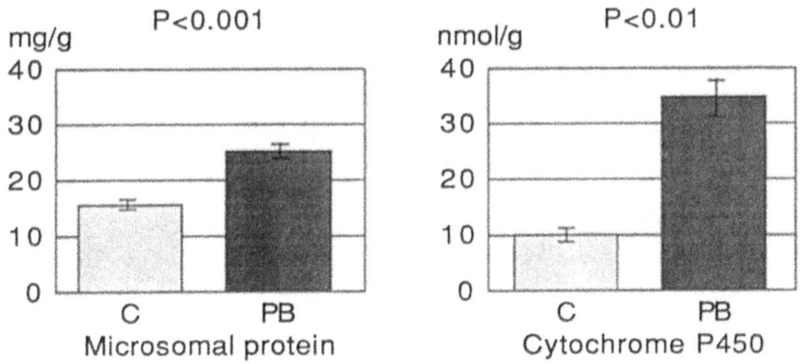

C = Control group PB = Phenobarbital treated group

FIGURE 3

HEPATIC MICROSOMAL ENZYMES AFTER 4 WEEKS

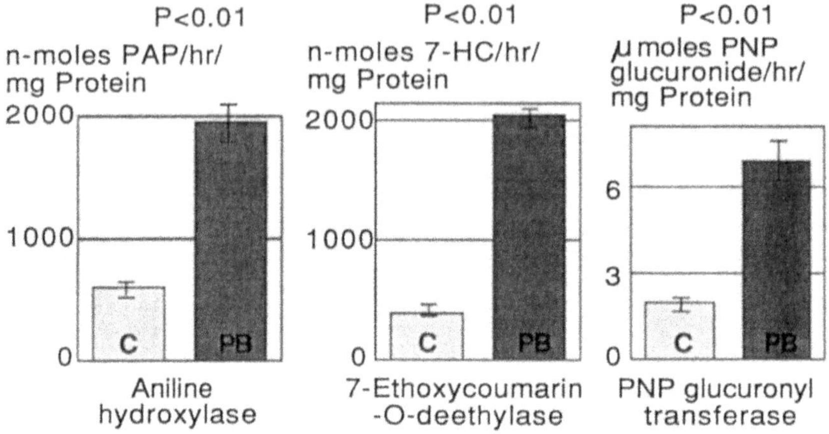

C = Control group PB = Phenobarbital treated group

FIGURE 4

PLASMA HORMONE LEVELS AFTER 4 WEEKS

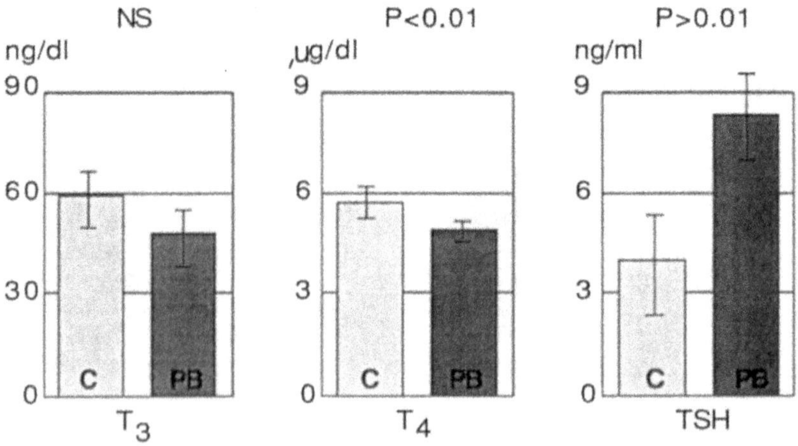

C = Control group PB = Phenobarbital treated group

In rodent studies TSH dependant growth progresses to malignancies with metastasis also. This is due to the aquisition of invasive properties prior to aquisition of TSH independant growth (Williams 1990). Long-term administration of compounds which results in TSH elevation induces hyperplasia and neoplasia including malignancies of follicular epithelium in the rat. Non-genotoxic compounds inducing follicular tumours mediated through TSH stimulation, should be evaluated by threshold effects rather than by the principles of genotoxic carcinogenicity (William 1990). As a rule, hormone dependant tumours are preceeded by a period of proliferative phase. Cell proliferation assays can be valuable to determine a threshold dose. It is possible to establish a threshold dose below which a safety margin can be based (Grasso, Sharratt and Cohen 1991).

When extrapolating results it is important to bear in mind that human thyroid is less sensitive to TSH than experimental animals. (Hill et al 1989). In familial goitre (inherited dishormonogenesis) although there is evidence of profound hyperplasia and nodular goitre, they seldom develop into carcinomas. This is in contrast to rat studies where hyperplasia often progress to adenomas and carcinomas.

2. ADRENERGIC SYSTEM

Certain broncho-dilators and cardiovascular drugs are known to influence the thyroid gland. Andrenergic system plays an important role in control of thyroid function and growth. Norepinephrine and certain α-2 agonists increase TSH secretion (Vijayan, Krulich, and McCann 1978). β-adrenergic mechanism is involved in the control of thyroid mitotic activity (Lewinski, Mroz-Wasilewska, Pawlikowski 1984). Thyroid gland contains adrenergic receptors which mediate nor-adrenaline induced iodide organification (Shimura, Endo and Onaya 1990).

3. LIVER

Hepatic thyroid axis is important to maintain balance between synthesis, metabolism and elimination of thyroid hormones. Hepatic microsomes possess enzymes linked with metabolism and elimination of thyroid hormones. Induction of hepatic microsomal drug metabolsing enzymes in rats not only alters xenobiotic handling but also changes the metabolism of thyroid hormones. Enzyme induction increases the conjugation and elimination of thyroxine which in turn result in lowering of plasma levels.

Phenobarbital (PB) was used as an example of an enzyme inducer and the following results were obtained to illustrate some of the above mentioned features. Administration of Phenobarbintone Sodium (75mg/kg/day) to young male rats (CD-Sprague-Dawley derived) for 4 weeks resulted in increased liver and thyroid weights (Fig.1). Hepatic microsomal protein, cytochrome P450, and enzymes such as aniline hydroxylase, ethoxy coumirine-O-deethylase and glucuronyl transferase showed significant increase when compared to controls (Figs.2 & 3).

Plasma levels of T3 and T4 showed reductions while serum TSH were significantly increased (Fig.4) when compared to controls. Histologically, there was centrilobular hepatocyte hypertrophy in rats given phenobarbital. Examination of thyroid gland showed follicular cell hypertrophy and hyperplasia with reduction of colloid. Morphometric analysis confirmed these findings (Figs. 5 & 6).

It is reported that phenobarbitone treatment increases biliary excretion of T4 and also increases biliary flow associated with induction of T4-glucuronyl transferase (McClean 1989). PB-treatment is also known to increase deiodination, and increased thyroxin binding in hepatocytes (Oppenheimer, Bernstein and Surks 1968). Thyroidal radio iodide accummulations are enhanced by certain compounds which increases thyroid clearance through hepatic mediation. These functional changes can be assessed by perchlorate discharge test (Alterwill, Collins, Brown, Harland 1987). The goitrogenic effect of PB can be prevented by T4 supplementation or hypophysectomy (McClean 1989). PB is shown to be a thyroid gland tumour promoter in initiation - promotion model (Hiasa, Kitahori, Oshima, Fujita, Yuasa, Konishi, Miyashiro 1982) when nitrosamine is used as an initiator. It is established that the tumour promotion is TSH mediated (McClain 1989) as hypophysectomy prevents the increased incidence of tumours in such a system.

In conclusion the follicular epithelium of rat is excessively sensitive to alterations in TSH levels. TSH elevation in rats rapidly leads to hypertrophy hyperplasia and neoplasia (including malignancies). Direct and indirect effects are involved in thyroid changes. It is important to investigate the role of mediation by other organ systems. Induced neoplasia of thyroid follicles in rats must be carefully extrapolated, from a safety evaluation point, as in majority of cases a threshold dose can be ascertained.

--- " ---

FIGURE 5
MORPHOMETRIC ANALYSIS

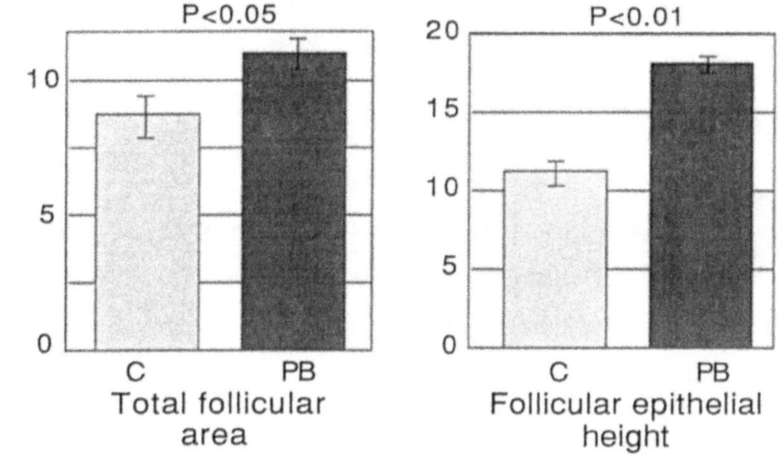

C = Control group PB = Phenobarbital treated group

FIGURE 6
MORPHOMETRIC ANALYSIS

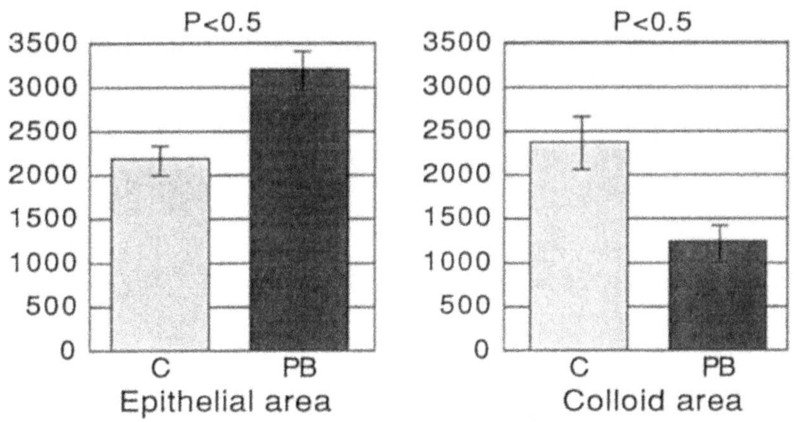

C = Control group PB = Phenobarbital treated group

TABLE 1

Classification of compounds acting on thyroid follicles.

1. GOITROGENS

   a. Natural/Dietary
      eg: cabbage, turnip

   b. Synthetic
      Thionamides - eg: Thiouracil, Carbamizole
      Aromatic amines - eg: Sulfa-drugs
      Polyhydric phenols - eg: Resorcinol

   Inhibits T3/T4 synthesis; Plasma T3/T4 reduced, TSH increased

2. ENZYME INDUCERS

   a. PB-type
      eg: DDT, Chlordane, Aldrin
   Hepatic microsomal enzymes and Cytochrome P450 increased.

   b. 3-Methyl Cholanthrene type
      eg: TCDD, PCB, PBB
   Hepatic microsomal enzymes and Cytochrome P450c increased.

   c. Mixed types
   T4-UDP Glucuronyl-transferase, biliary excretion of T4 and bile flow increased. Plasma T4 reduced, TSH increased.

3. INHIBITORS OF 5'MONODEIODINASE
      eg: Radio contrast agents, amiodarone
   Prevents peripheral conversion of T4 into T3. Plasma T4 increased, T3 decreased, TSH increased.

4. DIRECT ACTING CHEMICALS
      eg: N-Nitroso compounds
   Genotoxic - action not through pituitary axis

5. IONISING RADIATION
   Results in tumours in man and animals

TABLE 2

Thyroid Function Tests

1. Uptake of iodine by thyroid using ^{125}I

2. Perchlorate discharge test
   To assess iodide accumulation and organification mechanism

   (1 and 2 helps to differentiate between direct and indirect (peripheral) effects.)

3. Measurements of circulatory hormones
   eg: T3, T4, rT3 and TSH

4. Biliary excretion of ^{125}I-thyroxine and/or metabolites

5. Clearance of ^{125}I-thyroxine from plasma

6. Clearance of ^{125}I-triiodo thyronine (T3) from plasma

7. Measurement of hepatic microsomal drug metabolising enzymes
   eg: Glucuronyl transferase and Cytochrome P450

REFERENCES

Atterwill CK, Collin P, Brown CG and Harland RF (1987) The perchlorate discharge test for examining thyroid function in rats. J Pharmacol Methods 18:199-203

Grasso P, Sharratt M and Cohen AJ (1991) Role of persistent, non-genotoxic tissue damage in rodent cancer and relevance to humans. Annu Rev Pharmacol Toxicol 31: 253-287

Hiasa Y , Kitahori Y, Ohshima M, Fujita T, Yuosa T, Konishi N and Miyashiro A (1982) Promoting effects of phenobarbital and barbital on development of thyroid tumours in rats treated with N-bis (2-hydroxy propyl) nitrosamine. Carcinogenesis 3: 1187-1190

Hill RN, Erdreich LS, Paynter OE, Roberts PA, Rosenthal SL and Wilkinson CF (1989) Thyroid follicular cell carcinogenesis. Fund and Appl Toxicology 12: 629-697

Jemec B (1980) Studies of the goitrogenic and tumorigenic effect of two goitrogens in combination with hypophysectomy or thyroid hormone treatment. Cancer 45: 2138-2148

Kirschbaum A (1957) The role of hormones in cancer: Laboratory animals. Cancer Res 17: 432-453

Lewinski A, Mroz-Wasilewska Z and Pawlikowski M (1984) Effects of alpha-and beta-adrenergic agonists on the mitotic activity of thyroid follicular cells in vivo and in vitro. Endokrynologia Polska 35: 355-360

McClain RM (1989) The significance of hepatic microsomal enzyme induction and altered thyroid function in rats: Implications for thyroid gland neoplasia. Toxicologic Path 17: 294-305

Oppenheimer JH, Bernstein G and Surks MI (1968) Increased thyroxine turnover and thyroidal function after stimulation of hepatocellular binding of thyroxine by phenobarbital. J Clin Invest 47: 1399-1406

Rao RS, Heywood R and Gopinath C (in press) Age-related changes in thyroid structure and function in Sprague-Dawley rats. Vet Path

Shimura H, Endo T and Onaya T (1990) Thyrotrophin increases the α_{1b}-adrenergic receptors in rat thyroid gland in vivo. J of Endocrinol 126: 317-322

Vijayan E, Krulich L and McCann SM (1978) Catacholaminergic regulation in TSH and growth hormone release in overectemised and overectemised steroid primed rats. Neuro-endocrinology 26:174-185

Williams ED (1990) TSH and thyroid cancer. Horm Metabol Res Suppl 23: 72-75

MESULERGINE INDUCED LEYDIG CELL TUMOURS, A SYNDROME INVOLVING THE PITUITARY-TESTICULAR AXIS OF THE RAT

D E Prentice,* R A Siegel,** P Donatsch,** S Qureshi,** R A Ettlin**

*PCS, 7 Earning Street, Godmanchester, Cambs., UK
**Sandoz Pharma Ltd., CH-4002, Basle, Switzerland

INTRODUCTION.

The scientific literature abounds with reports of spontaneous and induced Leydig cell tumours in rats. The spontaneous lesions are age related and range from an incidence of 5% in Sprague-Dawley rats (Prejean 1973) to an incidence approaching 100% in Fischer 344 rats (Turek 1979). The older literature reports that disturbed hormone relationships and testicular damage were all important in the induction of Leydig cell tumours. Injections of cadmium salts and partial vascular ligation were frequently used to induce such tumours (Gunn 1965). Gunn in 1963 reported that levels of luteinising hormone were raised in rats treated three months previously with cadmium and Roe in 1965, speculated on the inter-relationships between the pituitary, the testes, the hormones which these two organs secrete and Leydig cell tumours.

More recently, widely-used pharmaceutical preparations from disparate drug classes have been reported to increase the frequency of testicular tumours in rats. The list includes:

- cimetidine, a histamine H_2-receptor antagonist (Leslie 1981; Brimblecombe 1984),
- hydralazine, an arteriolar vasodilator for hypertension (Griffith 1988),
- gemfibrozil, a lipid regulating agent (Griffith 1988),
- carbamazepine, an anticonvulsant and analgesic for trigeminal nerve neuralgia (Griffith 1988),
- vidarabine, an anti-viral agent (Griffith 1988)
- isradipine, a calcium channel antagonist for hypertension (Roberts 1989).

Mesulergine is an 8 alpha-aminoergoline which acts as an antagonist-agonist at D-2 receptors and was developed for the treatment of hyperprolactinaemia and of Parkinson's disease. During routine long-term carcinogenicity studies, the male Wistar rats developed an excess of Leydig cell tumours, some of which had an adenomatous pattern. This paper describes the explanatory studies which were performed to elucidate the mechanism of induction. The paper also compares the hormonal changes in the rat with mesulergine and those found with isradipine.

METHODOLOGY

Young (8 weeks at outset) or old (67 weeks at outset) male Wistar derived rats (Kfm:WIST) were used in eight studies whose duration ranged from acute single exposure studies to a two year study (see Table 1 below).

Mesulergine was dissolved in water and administered by gavage for the acute and two week studies. For studies of longer duration the compound was mixed in the feed. The dose levels chosen ranged from close to the intended human oral dosage to a maximum tolerated dose and all studies had a dose level equivalent to the high dose in the original long-term carcinogenicity study (1.6 mg/kg/day).

Table 1. List of Investigatory Studies

Study N°	Age at start(weeks)	Study Duration(weeks)	Doses(mg/kg/day)
1	8	acute	0.16, 1.6, 16.0
2	8	2	0.16, 1.6, 16.0
3	8	4	0.16, 1.6, 16.0
4	8	10	1.6
5	8	13	0.16, 1.6, 16.0
6	8	104	1.6
7	67	39	1.6

Blood samples were collected from the orbital sinus under light aether anaesthesia, immediately cooled on ice, centrifuged at 4°C and stored at -20°C. Serum levels of the following hormones were determined using specific RIA: luteinising hormone (LH), follicle stimulating hormone (FSH), testosterone (T), prolactin (PRL), growth hormone (GH), oestradiol (E2), adrenocorticotrophic hormone (ACTH), and corticosterone (CS).

Pathological examinations were undertaken in all studies (except studies 1 and 2) and consisted of full macroscopic post-mortem examination followed by weighing and microscopic examination of the testes. Leydig cell LH receptors were estimated in testicular material collected from studies 3, 5 and 7 (Hsueh 1978, 1979; Siegel 1991). The ligand was freshly prepared and purified ^{125}I-HCG. Frozen tissue samples were thawed and a crude membrane fraction prepared. Multiple dilution assays were performed.

The data relating to hormone levels are presented as arithmetic means with standard errors of the means and comparisons between drug treated groups and control groups were performed using the Student's T test. Testicular weights were compared using the Dunn test or Dunnett's test (Miller 1966; Dunnett 1955).

RESULTS.

Study 1 (acute): Serum LH and T levels were dose-dependently reduced by mesulergine and PRL concentrations were maximally suppressed by all doses. The highest dose of mesulergine also led to changes in circulating levels of GH (decrease), ACTH and CS (increases) (hormone data not presented). No effects on serum E2 concentrations were observed.

Study 2 (2 week): No consistent or dose-dependent effects of mesulergine on any hormone were observed with the exception of reduced serum PRL levels (hormone data not presented).

Study 3 (4 week): Serum LH levels were dose-dependently elevated and the effect attained statistical significance at the highest dose level (see Table 2). The drug also reduced circulating PRL levels and at the highest dose, elevated ACTH concentrations. No drug effects upon T, E2, or GH were observed. Leydig cell LH receptor concentrations, as expressed by ^{125}I-HCG binding, were reduced by all doses of mesulergine (data not presented). Pathological investigations revealed no changes which could be attributed to treatment.

Table 2. Study 3, Serum Hormone Levels

Hormone	Control	Low dose	Mid dose	High dose
LH (ng/ml)	78 ± 9	109 ± 11	124 ± 21	154 ± 18[2]
T (pg/ml)	1514 ± 350	1402 ± 338	1075 ± 339	1016 ± 168
E2 (pg/ml)	13 ± 1.0	13 ± 0.4	12 ± 1.02	12 ± 1.2
PRL (ng/ml)	4.3 ± 1.4	2.1 ± 0.7	1.2 ± 0.1	2.0 ± 0.9
GH (ng/ml)	298 ± 97	234 ± 92	129 ± 80	154 ± 75
ACTH (pg/ml)	41 ± 6	45 ± 6	43 ± 3	208 ± 83[1]

[n/group = 6, [1] = P<0.05, [2] = <0.01]

<u>Study 4 (10 weeks)</u>: Serum LH levels were significantly elevated, and those of PRL were reduced in the mesulergine treated rats. No effects on circulating T concentrations were detected, nor were there any alterations in testicular weight or morphology (hormone data not presented).

<u>Study 5 (13 weeks)</u>: Serum LH and ACTH were significantly elevated by mesulergine at 16 mg/kg/day, while PRL secretion was inhibited at all dose levels (see Table 3). As in the 4 week study, Leydig cell receptor levels were reduced by mesulergine and the effect was significant with the mid and high dose animals (data not presented). Again pathological investigations failed to show treatment-related changes.

Table 3. Study 5, Serum Hormone Levels

Hormone	Control	Low dose	Mid dose	High dose
LH (ng/ml)	47 ± 5	58 ± 13	54 ± 10	74 ± 7[2]
T (pg/ml)	706 ± 262	668 ± 212	553 ± 157	1104 ± 310
E2 (pg/ml)	18.2 ± 3.7	19.0 ± 0.6	20.0 ± 4.0	18.3 ± 1.4
PRL (ng/ml)	12.2 ± 0.9	2.2 ±0.4[3]	2.6 ± 0.2[3]	2.8 ± 0.3[3]
GH (ng/ml)	88 ± 36	303 ± 97	30 ± 11	270 ± 95
ACTH (pg/ml)	47 ± 11	185 ± 77	412 ± 210	924 ± 250[1]

[n/group = 6, [1] = P<0.05, [2] = <0.01, [3] = P<0.001]

<u>Study 6 (104 weeks)</u>: Mesulergine treatment led to elevated serum LH levels throughout most of the lifetime of the animals (see Table 4). The effect was marked and statistically significant at all time points up to week 79: from that time onwards, the LH secretion stimulating effects of the drug became less marked and were no longer statistically significant. Serum FSH levels were elevated on most occasions, but the change attained statistical significance at one time point only, namely week 39.

As expected, circulating PRL concentrations were markedly reduced by mesulergine throughout the study. In spite of the consistent drug-induced changes in LH and PRL levels, there was no effect on blood T levels (data not shown).

Histopathological examination shows that mesulergine treatment was associated with Leydig cell tumours and hyperplasias in 77% of rats compared to 8% in controls (n = 12/group).

Table 4. Study 6, Serum LH, FSH and PRL Concentrations

Hormone	Duration of treatment (weeks)	Control	Treated
LH (ng/ml)	14	77 ± 7	168 ± 25[2]
	27	93 ± 5	233 ± 33[3]
	39	147 ± 10	280 ± 28[3]
	53	77 ± 9	164 ± 11[3]
	66	93 ± 8	164 ± 24[1]
	79	160 ± 30	211 ± 33
	92	174 ± 29	164 ± 30
	105	65 ± 7	106 ± 18
FSH (ng/ml)	14	NA	NA
	27	NA	NA
	39	8.8 ± 0.8	12.0 ± 1.2[1]
	53	15.5 ± 2.1	16.4 ± 1.3
	66	7.8 ± 1.6	11.2 ± 1.1
	79	21 ± 1.3	19 ± 1.4
	92	17 ± 3.1	24.5 ± 4.0
	105	13 ± 2.7	13 ± 2.9
PRL (ng/ml)	14	143 ± 6	4.9 ± 0.2[3]
	27	44 ± 5	1.2 ± 0.1[3]
	39	>100	9.3 ± 0.8[3]
	53	96 ± 14	2.2 ± 0.2[3]
	66	NA	NA
	79	87 ± 17	2.0 ± 0.4[3]
	92	NA	NA
	105	NA	NA

[n/group = 10, [1] = P<0.05, [2] = P<0.01, [3] = P<0.001, NA = not available]

<u>Study 7 (39 weeks, aged rats)</u>: At both 13 and 26 weeks, significant elevations in serum LH levels, along with marked reduction in circulating PRL concentrations, were observed in treated rats as compared to controls (see Table 5). Following 39 weeks of treatment (rats aged 106 weeks), only a slight and statistically not significant increase in LH levels was apparent. No drug effects on serum T were seen at any time point. Leydig cell LH receptor concentrations were markedly reduced in treated rats compared to controls (data not presented). In contrast to the 104 week study, histological examination and testicular weight analysis did not reveal any treatment related changes.

Table 5. Study 7, Serum LH, T and PRL Concentrations

Hormone	Duration of treatment (weeks)	Control	Treated
LH (ng/ml)	13	60 ± 4	140 ± 17^{2}
	26	64 ± 5	91 ± 7^{2}
	39	75 ± 10	95 ± 15
T (ng/ml)	13	849 ± 118	785 ± 101
	26	966 ± 136	871 ± 139
	39	NA	NA
PRL (ng/ml)	13	85 ± 12	0.8 ± 0.1^{3}
	26	44 ± 5	3.7 ± 0.4^{2}
	39	NA	NA

[n/group = 10-17, ^{2} = P<0.01, ^{3} = P<0.001, NA = not available]

DISCUSSION

The present experiments demonstrate that in the rat mesulergine, in addition to inhibiting PRL release, exerts pronounced effects upon LH secretion. These LH-modulating effects appear to be a function of the duration of application of the compound. In the acute studies mesulergine led to reduced serum levels of LH. The transient nature of this inhibitory action on the hypothalamo-hypophyseal-gonadal (HHG) axis was illustrated by the two week studies, in which no drug-related effects on LH or T were apparent. Subsequently, in the various longer-term studies, chronically elevated serum LH levels were seen. The stimulatory effect upon LH secretion has been established in young animals during almost their entire lifetime and in aged rats (67 weeks old at the beginning of the study) following 13 and 26 weeks of treatment with mesulergine. Thus, these long-term studies demonstrate that in male rats, continuous exposure to mesulergine, in the dose range capable of increasing the incidence of Leydig cell tumours, induces chronically elevated circulating LH levels. This effect first manifests itself following approximately one month of drug administration.

LH comprises the major hormonal input to Leydig cells, altering their morphology as well as stimulating their secretory function (Hodgson 1983), and there exists considerable experimental data linking LH to Leydig cell hyperplasia and tumours in sub-human species. Increased volume of Leydig cell clusters in adult male rats treated with gonadotrophic hormones was first demonstrated over fifty years ago (Greep 1936, 1937). This hyperplasia- and hypertrophy-inducing effect of exogenous LH and human chorionic gonadotropin (HCG) has since been amply confirmed in primates as well as in other rodent species (Kerr 1986; Christensen 1980; Schoen 1964; Pfeiffer 1943; Simpson 1954).

In addition to exogenous hormone, other studies have also implicated endogenous LH in the development of Leydig cell hyperplasia and tumours in rats. A correlation has been demonstrated between high pituitary LH concentrations and tumours, while suppression of LH synthesis prevented the development of the tumours (Lee 1981). In a long-term (20 month) model in which male rats were parabiosed to castrated animals, the data suggested that prolonged elevations in circulating levels of endogenous gonadotropins correlated with an increased incidence of Leydig cell tumours (Brown 1979).

This clear evidence for gonadotropin-related Leydig cell hyperplasia and tumours, leads us to believe that the sustained elevated circulating LH levels which we have observed in animals treated chronically with mesulergine are the direct cause of the increased incidence of Leydig cell tumours seen in the carcinogenicity study.

In the rat carcinogenicity study mesulergine had no apparent effect on the time of tumour appearance. This is somewhat puzzling, since our present studies have shown that elevations in serum LH levels can be detected as early as four weeks following the onset of treatment. An explanation for this apparent paradox may be found in the fact that the spontaneous development of testicular tumours in rats is an age-related phenomenon in the normal (drug-free) endocrine milieu. That is to say, old age is a prerequisite for tumour occurrence, even in those rat strains in which the spontaneous incidence is high. Administration of exogenous gonadotrophic hormones can result in Leydig cell tumour formation within a time period of weeks, but this is probably due to the exceedingly high serum hormone levels achieved in these experiments.

The present experiments have underlined the expected marked and sustained suppressive effect of mesulergine on PRL secretion. Therefore, the possible direct involvement of PRL in the Leydig cell hyperplasia/tumour phenomenon must be considered. PRL is not, however, known to exert trophic effects upon the testes, its action in this gland being the regulation of Leydig cell LH receptors. In the rat, Leydig cell LH receptor number is a function of circulating PRL levels (Huhtaniemi 1981; Bartke 1976; Zipf 1978; Purvis 1979). It was therefore considered it likely that chronic mesulergine administration might result in reduced Leydig cell LH receptor function. Our present findings support this supposition. We have observed reduced LH binding capacity in testicular material from rats treated for 4 or 13 weeeks with a wide range of doses of mesulergine. This observation was also made in aged rats treated for 39 weeks with the drug. This produced an imbalance in the HHG axis (a closed-loop system), whereby reduced Leydig cell responsiveness to LH would be compensated for by chronically elevated LH secretion resulting in normal circulating levels of T. The present experimental findings support the hypothesis that mesulergine-induced Leydig cell tumours in the rat carcinogenicity study were caused by chronically elevated circulating LH levels.

The investigatory studies with isradipine (Roberts 1989) also showed that gonadotropins were involved in Leydig cell tumour induction. In comparison with mesulergine, there were many similarities, i.e. acutely, serum T levels were depressed, the gonadotropin modulatory effects required long-term administration of the drug and levels of Leydig cell LH receptors were reduced. There were, however, some important differences; isradipine clearly elevated both gonadotropins, with FSH levels attaining significance at several time points. Gonadotropin elevations were detected only after 52 weeks of treatment (4 weeks for mesulergine). Testicular T levels were raised after 10 weeks administration (not raised with mesulergine) and there was only a mild reduction of circulating PRL (compared to pronounced reduction with mesulergine). Interestingly, cimetidine also leads to elevated serum LH levels in the rat (Baba 1981).

Chronic administration of compounds which cause elevations of circulating LH in rats, ultimately induces Leydig cell tumours and the syndrome appears to be rat-specific. Similar changes are not found in other laboratory animal species. This syndrome may arise from any perturbation of the closed-loop feedback mechanism of the HHG axis. With mesulergine, the peturbation was initiated by reduction of circulatory PRL and an associated reduction in Leydig cell LH receptor concentrations. Although the precise cause of the perturbation may vary between compounds, the end result appears to be the same in each case.

REFERENCES

Baba S, Paul HJ, Pollow K, Jaretschek G, Jacobi GH (1981) In vivo studies on the antiandrogenic effects of cimetidine versus cyproterone acetate in rats. The Prostate 2: 163-174

Bartke A (1976) Pituitary testes relationships: role of prolactin in the regulation of testicular function. Prog Reprod Biol 1: 136-150

Brimblecombe RW, Leslie GB (1984) Cimetidine. IN: Laurence DR, McLean AEM, Weatherall M (eds) Safety testing of new drugs. Academic Press, London, pp 65-91

Brown CE, Warren S, Chute RN, Ryan KJ, Todd RB (1979) Hormonally induced tumors of the reproductive system of parabiosed rats. Cancer Res 39: 3971-3976

Christensen AK, Peacock KC (1980) Increase in Leydig cell number in testes of adult rats treated chronically with an excess of human chorionic gonadotrophin. Biol of Reproduction 22: 383-391

Dunnett ChW (1955) A multiple comparison procedure for comparing several treatments with a control. J Am Stat Ass 50: 1096-1121

Greep RO, Fevold HL, Hsaw FL (1936) Effect of two hypophyseal gonadotrophic hormones on the reproductive system of the male rat. Anat Rec 65: 261-271

Greep RO (1937) Pituitary regulation of the male gonad. Cold Spring Harbor Symp Quart Biol 5: 136-143

Griffith RW (1988) Carcinogenic potential of marketed drugs. J Clin Res and Drug Develop 2: 141-144

Gunn SA, Gould TC, Anderson WAD (1963) The selective injurious response of testicular and epididymal blood vessels to cadmium and its prevention by Zinc. Amer J Path 42: 685-702

Gunn SA, Gould TC, Anderson WAD (1965) Comparative study of interstitial cell tumours of rat testis induced by cadmium injection and vascular ligation. J Natl C Instit 35: 329-337

Hodgson Y, Hudson B (1983) Leydig cell function. Monogr on Endocrinology 7: 107-132

Hsueh AJW, Dufau ML, Catt KJ (1978) Direct inhibitory effect of estrogen on Leydig cell function of hypophysectomised rats. Endocrinol 103: 1096-1102

Hsueh AJW, Erickson GF (1979) Extra-pituitary inhibition of testicular function by luteinizing hormone releasing hormone. Nature 281: 66-67

Huhtaniemi IT, Catt KJ (1981) Induction and maintenance of gonadotrophin and lactogen receptors in hypoprolactinemic rats. Endocrinol 109: 483-490

Kerr JB, Sharpe RM (1986) Effects and interactions of LH and LHRH agonist on testicular morphology and function in hypophysectomized rats. J Reprod Fert 76: 175-192

Lee Ch., Jesik Ch., Falkowski W, Uke E, Grayhack JT (1981) Testicular neoplasia: Clinical and experimental. In: Nagasawa H, Abo K (eds) Hormone related tumours, pp 289-311

Leslie GB, Noakes DN, Pollitt FD, Roe FJC, Walker TF (1981) A two-year study with cimetidine in the rat: assessment for chronic toxicity and carcinogencity. Toxicol and Applied Pharmacol 61: 119-137

Miller RG (1966) Simultaneous statistical inference. McGraw Hill, New York

Pfeiffer CA, Hooker ChW (1943) Testicular changes resembling early stages in the development of interstitial cell tumours in mice of the A strain after long-continued injections of pregnant mare serum. Cancer Res 3: 762-766

Prejean JD et al. (1973) Spontaneous tumors in Sprague-Dawley rats and Swiss mice. Cancer 33: 2768-2773

Purvis K, Clausen DPF, Olsen A, Hang E, Hansson V (1979) Prolactin and Leydig cell responsiveness to LH/HCG in the rat. Arch Androl 3: 219

Roberts SA, Nett TM, Hartman HA, Adams TE, Stoll RE ((1989) SDZ 200-100 induces Leydig cell tumors by increasing gonadotrophins in rats. J Amer Coll of Toxicol 8: 487-505

Roe FJC, Dukes CE, Cameron KM, Pugh RCD, Mitchley BCV (1964) Cadmium neoplasia: testicular atrophy and Leydig cell hyperplasia and neoplasia in rats and mice following the subcutaneous injection of cadmium salts. Brit J Cancer 17: 674-681

Schoen EJ (1964) Effect of local irradiation on testicular androgen biosynthesis. Endocrinol. 75: 56-65

Siegel RA, Hsueh AJW, Donatsch P, Marko M, Flueckiger E, Prentice DE (1991) Effects of long-term treatment with the dopaminomimetic mesulergine (CU 32-085) on Leydig cell LH receptors, and on Leydig cell responsiveness to LH, in rats and mice. In preparation.

Simpson ME, van Wagenen G (1954) Persistent nodules in testis of the monkey associated with Leydig cell hyperplasia induced by gonadotrophins. Cancer Res 14: 289-293

Turek FW, Desjardins C (1979) Development of Leydig cell tumors and onset of changes in the reproductive and endocrine systems of aging Fischer F344 rats. J Natl C Instit 63: 969-975

Zipf WB, Payne AH, Kelch RG (1978) Prolactin, growth hormone and LH in the maintenance of testicular LH receptors. Endocrinol 103: 595

VARIATIONS IN PREVALENCE OF ENDOCRINE TUMORS AMONG DIFFERENT COLONIES OF RATS ? A RETROSPECTIVE STUDY IN THE HANNOVER REGISTRY DATA BASE

G. Morawietz and S. Rittinghausen

Fraunhofer Institute of Toxicology and Aerosol Research, Nikolai-Fuchs-Str. 1, DW-3000 Hannover 61, Germany

INTRODUCTION

Long-term carcinogenicity and toxicity studies in laboratory rodents are crucial steps in the safety evaluation of drugs and chemicals. All these studies include at least one control group required for the correct interpretation of findings. In order to evaluate and interpret the results of these studies by comparing the incidence and degree of pathological alterations in treatment groups with those in control groups, a fundamental knowledge of the frequency and type of spontaneously occurring lesions (in particular tumors) is necessary, taking into consideration differences between animal species and strains (Ettlin *et al.* 1991).

The source of information used for comparison and interpretation of animal experiments is primarily the concurrent control group of a particular study. In doubtful or difficult cases, for example due to a limited number of animals in a control group or because of an unexpectedly high or low spontaneous tumor rate in a particular organ, other sources of information are needed to interpret the results of statistical analysis. If available, in-house historical control data of previous studies, which have been carried out with the same animal strain under comparable experimental conditions, are used for this purpose. Otherwise, data published in literature or given by the breeder have to be used for argumentation.

A problem occurring here is, that the diagnostic criteria used by pathologists and/or given in literature are sometimes ambiguously defined, which may cause confusion in the interpretation of study data (Haseman *et al.* 1984). Although several nomenclature proposals and guidelines had been published in the past, presenting either a structured system of names and codes for organs and lesions or defining diagnostic criteria to be used in animal pathology, a *standardized* and *systematized* nomenclature system for experimental animal pathology, covering all organ systems and suitable for use in electronic data processing, was still lacking.

In 1988 the Hannover REGISTRY Data Base was set up in co-operation with 13 chemical and pharmaceutical companies[1] from Germany and Switzerland with the purpose of establishing a large pool of historical control data. Because such a project could only be successful if standardized diagnostic criteria are applied, this system represents an integration of two major parts:

- a *data base* for the collection of spontaneously occurring pre-neoplastic and neoplastic lesions in laboratory animals

  and

- a *systematized nomenclature* system for laboratory rodents (currently only rats) which is designed to be suitable for electronic data processing.

The simultaneous development of both systems by the same group of pathologists and computer experts guarantees that diagnoses collected from different studies and stored in this system are *comparable* because the underlying diagnostic criteria are standardized.

DATA BASE CONTENTS

Histopathological findings form the major part of the information to be stored in a pathology data base. But as the frequency and occurrence of spontaneous lesions in laboratory animals are influenced by various factors, such as the animal strain or maintenance and environmental conditions, detailed information on these topics is also stored in the data base beside the actual histopathological diagnoses (Morawietz and Rittinghausen 1990). Three main tables are used to store study-related data, animal-related data and histopathological diagnoses.

Study-related data describe the experimental conditions of an animal study, namely the duration of an experiment, housing and environmental data, such as the temperature in the animal room and other physical parameters, information about the diet used as well as data on the bedding material and a possible treatment or sham treatment of the control group. All in all, more than 50 single data items per study are recorded in the data base.

1 ASTA Pharma AG, BASF Aktiengesellschaft, BAYER AG, Boehringer Ingelheim KG, Boehringer Mannheim GmbH, Byk Gulden Pharmazeutika, CIBA-GEIGY AG, HOECHST AG, F. Hoffmann-LaRoche AG, KNOLL AG, E. MERCK, Dr. Karl THOMAE GmbH, SANDOZ Pharma AG

Another type of important information is the animal-related data. While the study-related information describes the general conditions and the design of a study, the animal-related data are specific to each individual rat used in the control group. Recorded in the animal table are the species (currently rats), the strain, information on the breeder, the sex of the individual animal, the age of a rat at the start and end of the study, the animal status, and many other parameters.

The most important information to be stored in the data base is of course the histopathological diagnoses. Primarily, the localization of a lesion (the organ) and the morphological term (the name of the lesion) have to be stored. Furthermore, a diagnosis can be extended by one or more modifiers for a more precise subclassification, for example, to specify a specific growth pattern or a distinct cell type. The item "biological behavior" is used to declare the malignancy of a tumor and in particular to define whether a malignant tumor is metastasizing or not. A possible multiplicity of a lesion and its unilateral or bilateral occurrence in paired organs is stored in the data base as well as information about the histological methods applied.

THE REGISTRY NOMENCLATURE

In order to describe a histopathological diagnosis in the way already mentioned and to store all the data in a computerized data base, distinct definitions and rules have to be observed for each single data item to be able to evaluate all the data and to produce reports.

The criteria for a histopathological diagnosis are described in many text books and publications, namely in the nomenclature of the World Health Organization (Turusov and Mohr 1990) or in the ILSI monographs (Jones *et al.* 1983, 1985, 1986, 1987, 1988, 1989, 1990). But in this literature on laboratory animals, different names are often used for lesions of the same histological type and the criteria which lead to a specific diagnosis are sometimes ambiguously defined. Therefore, the REGISTRY group started to define standardized criteria for the diagnoses of neoplastic and pre-neoplastic findings in the rat.

Although the REGISTRY nomenclature is primarily based on the existing and approved literature, in several cases the system was especially adapted for use in practical toxicological pathology. Complying with the major rule, to be able to use the nomenclature within a computerized data base, the first step was to set up a systematized structure for organs and lesions (Mohr *et al.* 1990).

Standardization of names for organs and lesions and building a hierarchic structure is the first but just one prerequisite for a systematized nomenclature used to operate a pathology data base. Another even more important point is the agreement on standardized diagnostic criteria used to characterize a specific lesion (Morawietz *et al.* 1990). Therefore, detailed descriptions and definitions for all tumors and pre-neoplastic lesions in all organ systems have been worked out by the team of pathologists participating in the REGISTRY project. Up to now this nomenclature has been prepared for rats but it is anticipated to extend the definitions to mice in the near future.

Although the project is entitled *Tumor* REGISTRY data base, relevant pre-neoplastic lesions are also included in the REGISTRY nomenclature and appropriate diagnoses are stored in the data base. An exact definition of the differentiation between pre-neoplastic lesions and tumors is mandatory to facilitate producing reports on the *tumor* incidence. Similar reports can also be generated for pre-neoplastic lesions if their differentiation from non-neoplastic diseases is defined in an analogous way.

The REGISTRY nomenclature for the rat currently consists of about 300 manuscripts, which will be published under the supervision of WHO/IARC (International Agency for Research on Cancer) and ILSI (International Life Sciences Institute) in the near future.

Another benefit of this nomenclature system is that all definitions and criteria are not only theoretically defined. They are also practically approved in many slide sessions and panel meetings held with the group of participating pathologists. During the development of the system several thousand histological slides of finalized studies have been reviewed.

The panel meetings currently take place about every second month to validate the data of new studies to be entered into the data base in order to assure that each individual diagnosis is consistent with the defined nomenclature. During the validation procedure all findings of a study are pre-checked by pathologists at the Fraunhofer Institute and selected cases by a panel of pathologists from the participating companies. The findings which are reviewed by the panel are questionable diagnoses or borderline cases and interesting or rare lesions. This meticulous procedure guarantees that future evaluations of the collected data will yield reliable and consistent results. An additional benefit is of course continuous training of the participating pathologists.

FIRST RESULTS

The purpose of the examples presented here is *NOT* to present novel results on the tumor spectrum or tumor frequency in distinct organs, because most of the findings are already described in literature. It is to demonstrate the influence of parameters like sex, strain, age etc. on the frequency of spontaneously occurring tumors and to compare the results obtained from other authors with the information from the data base. Precise knowledge of possible effects of such parameters is important when using historical control data for a comparison or interpretation of animal studies (Haseman *et al.* 1984). This applies to information derived from literature as well as for a computerized data base.

As mentioned earlier, more than 50 single data items are stored in the data base to describe the study design and environmental conditions of an animal experiment and in addition several parameters on each individual animal. The amount of data which is currently available from the data base is much too small to find out detailed relationships of those parameters. In total, there are data from about 3000 control rats in the data base. All these animals were involved in 24 long-term carcinogenicity or toxicity studies. To detect the influence of subtle effects on the frequency of spontaneously occurring lesions, especially when looking for rare tumors, more data are necessary.

The frequency of spontaneous tumors in various organs of the endocrine system (pituitary gland, thyroid glands, parathyroid glands, adrenal glands, and endocrine pancreas) was determined for Wistar and Sprague-Dawley rats separately.

An evaluation on the total number of animals bearing endocrine tumors amounts in Wistar rats to a frequency of 47 % (464/990) in males and 63 % (619/990) in females. The Sprague-Dawley strain shows a frequency of 64 % (305/480) in males and 67 % (320/480) in females. In both strains, the incidence for females is higher than for males and the difference between the sexes is more pronounced in the Wistar strain. Relatively high incidences of endocrine tumors in the same order of magnitude were also examined for the Wistar strain by Bomhard *et al.* (1986), and Deerberg *et al.* (1980), and for the Sprague-Dawley rats by Suzuki *et al.* (1979), Löhrke *et al.* (1982), and Kaspareit-Rittinghausen *et al.* (1990a).

Another important question during the evaluation of animal studies is the multiplicity of tumors of different histogenesis. Figure 1 shows the results from the data base, again for tumors of the endocrine system. Plotted here are the frequencies for animals with one, two, and more than two endocrine tumors of different origin, again for both sexes of Wistar and Sprague-Dawley rats.

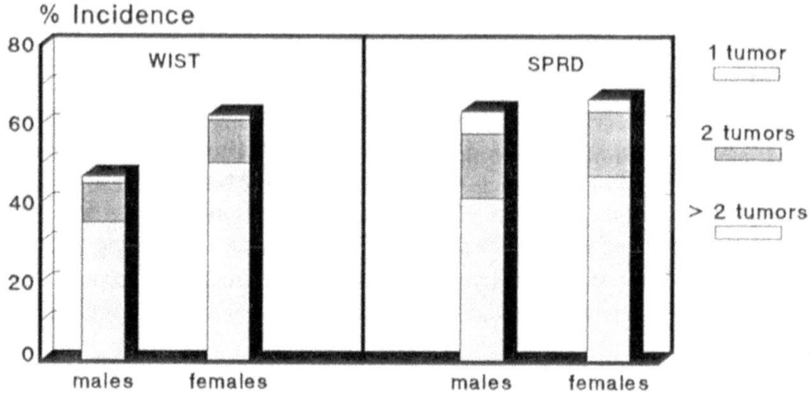

Fig. 1. Multiplicity of endocrine tumors of different histogenesis for Wistar and Sprague-Dawley rats

Aged Sprague-Dawley rats develop more multiple tumors than aged Wistar rats. In the Wistar strain, about 10 % of the animals (males 97/990, females 108/990) have 2 different endocrine tumors. The Sprague-Dawley strain shows a value which is about twice as high (males 79/480, females 79/480). Looking for the number of animals with more than 2 endocrine tumors, the difference between both strains is even more pronounced: only 2 % (18/990) of the male Wistar rats and 1 % (11/990) of the female Wistar rats have more than 2 tumors compared to about 6% (27/480) of the male Sprague-Dawley rats and 3 % (15/480) of the female Sprague-Dawley rats. The maximum number of endocrine tumors observed in a single animal was 4. Similar results were observed by Suzuki *et al.* (1979).

The most common tumors occurring in organs of the endocrine system, as reported in literature, are benign C-cell tumors of the thyroid gland (Kaspareit-Rittinghausen *et al.* 1990b), adenomas of the pars distalis of the pituitary gland (Bomhard *et al.* 1986) and benign medullary tumors of the adrenal glands (Löhrke *et al.* 1982). Comparing the results of different authors there is a considerably high variation in tumor rate and tumor spectrum even for the same strain of rats. In the REGISTRY data base, the most common endocrine tumor for both strains in all breeding colonies is the pars distalis adenoma of the pituitary gland. The benign C-cell tumor of the thyroid glands and the benign medullary tumor of the adrenal glands are the second frequently observed neoplasms in Wistar rats. In Sprague-Dawley rats cortical adenomas were more often diagnosed than medullary tumors.

The animals used in the reported studies were obtained from different breeding colonies. In Fig. 2 the data from endocrine tumors which are currently stored in the REG-

ISTRY data base are assigned to 4 different breeders of Wistar rats (denoted as A, B, C, and D) and to 2 different breeders of Sprague-Dawley rats (denoted as E and F, respectively). The plot summarizes the frequencies from multiple studies for each of the 6 breeding colonies and shows that the female animals in most of the sub-strains develop a higher incidence of endocrine tumors than the males.

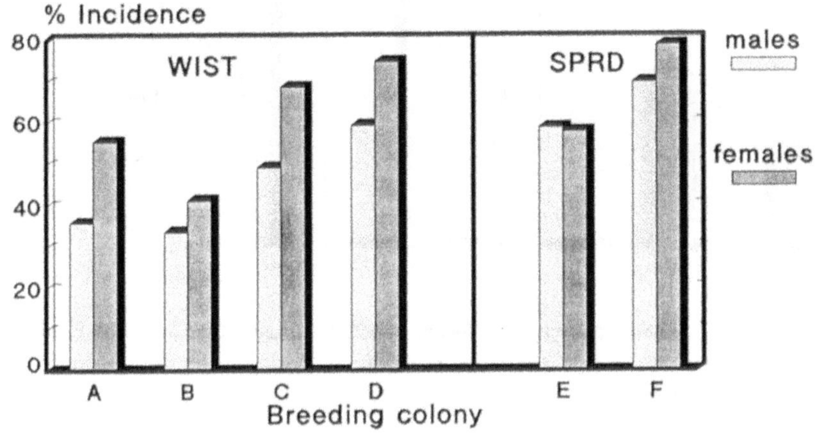

Fig. 2. Frequency of spontaneously occurring endocrine tumors observed in animals of different breeding colonies

The apparent variation between different breeding colonies may have been caused by many reasons. One reason could be the age of the animals, because the data shown in Fig. 2 result from long-term studies, which were terminated after between 24 and 31 months and it is well known that increasing age has a strong influence on the spontaneous tumor rate (Anver *et al.* 1982).

An example for the dependency of the tumor rate on the age of the animals is given in Fig. 3. The data plotted here were derived only from the breeding colony B, and again the total amount of tumors observed in organs of the endocrine system is shown. The columns on the left side represent the results from a 24-month study and on the right side the corresponding values from a 30-month study are shown. Not only the difference in the total tumor rate is remarkable but also the fact that 9 % of the older males and 5 % of the older females have more than 1 tumor, compared to only 1 % of the younger males and 0 % of the younger females. For this evaluation, only the nominal duration of the studies was taken into account. With respect to the data base, it is also possible to define arbitrary time windows and to consider separately the animals which died or were killed in moribund condition. Such detailed analysis of all parameters possibly accounting for the observed differences will be essential for establishing genuine causal relationships.

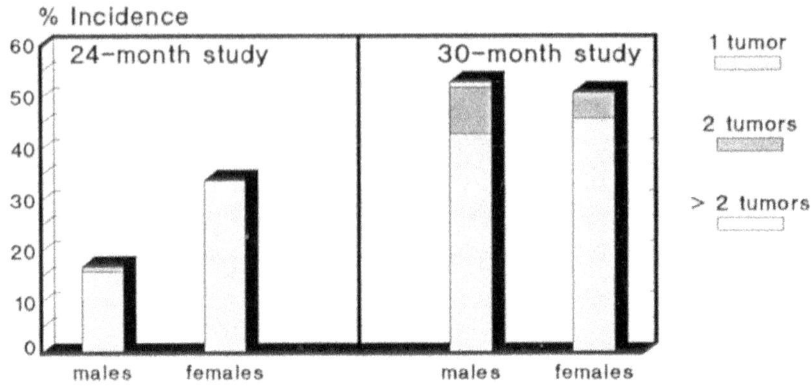

Fig. 3. Dependency of the rate of spontaneous endocrine tumors on the age of the animals derived from one breeding colony (B)

SUMMARY AND CONCLUSION

Illustrated by data derived from the endocrine system it could be demonstrated that at least the following requirements have to be fulfilled when using historical control data for the interpretation of animal studies:

- The main and basic requirement is the utilization of standardized diagnostic criteria and a systematized nomenclature for all data which should be compared. This includes the historical control data as well as histopathological diagnoses obtained from other animal studies. All findings stored in the REGISTRY are based on standardized criteria and this nomenclature was also applied for the more recent studies which were evaluated by pathologists of the member companies. Data from literature which are frequently used for comparison and interpretation cannot always fulfil this requirement due to the variation of diagnostic criteria used by different investigators.

- The second important requirement is an exact knowledge of study-related details. Some dependencies of the tumor rate on the sex, the strain, the breeder and the age were shown in a few examples. But other parameters, like diet, caging, bedding etc. may also influence the frequency of spontaneously occurring tumors and therefore have to be taken into account. In the REGISTRY such data from routine carcinogenicity and toxicity studies are available because all these studies are performed under GLP conditions and all parameters are carefully recorded.

During the relatively short period of its activity, the REGISTRY has become not only source of historical laboratory data, but also it subserves as indispensable tool of quality control warranted by peer-review effects of panel meetings.

An additional benefit will emanate from the data base in the near future: Exact knowledge of the magnitude of the influence of distinct parameters will enable the member companies to use the information for the design or optimization of new studies, as well as to extend the analysis and support the interpretation of observed results.

Acknowledements: The authors thank R. Bader, U. Deschl, H. Ernst, R. Ettlin, C. Gembardt, S. Halm, J.H. Harleman, F. Hartig, E. Hartmann, K. Heider, W. Jahn, E. Karbe, W. Kaufmann, K. Krieg, G. Krinke, K. Küttler, C. Landes, F. Mettler, U. Mohr, H. Püschner, S. Qureshi, G. Reznik, K. Tuch, H. Urwyler, G. Weisse, and I. Weisse for their co-operation in the REGISTRY project.

REFERENCES

Anver MR, Cohen BJ, Lattuada CP, Foster SJ (1982) Age-associated lesions in barrier-reared male Sprague-Dawley rats: A comparison between Hap:(SD) and Crl:COBS CD (SD) stocks. Exp Aging Res 8: 3-24

Bomhard E, Karbe E, Loeser E (1986) Spontaneous tumors of 2000 Wistar TNO/W.70 rats in two-year carcinogenicity studies. J Environ Pathol Toxicol Oncol 7: 35-52

Deerberg F, Rapp KG, Pittermann W, Rehm S (1980) Zum Tumorspektrum der Han:WIST-Ratte. Z Versuchstierk 22: 267-280

Ettlin RA, Oberholzer M, Perentes E, Ryffel B, Kolopp M, Qureshi SR (1991) A brief review of modern toxicologic pathology in regulatory and explanatory toxicity studies of chemicals. Arch Toxicol 65: 445-453

Haseman JK, Huff J, Boorman GA (1984) Use of historical control data in carcinogenicity studies in rodents. Toxicol Pathol 12: 126-135

Jones TC, Mohr U, Hunt RD (eds) (1983, 1985, 1986, 1987, 1988, 1989, 1990) Monographs on pathology of laboratory animals, Vol 1-7. Springer, Berlin Heidelberg New York Tokyo

Kaspareit-Rittinghausen J, Deerberg F, Rapp K, Wcislo A (1990a) Mortality and tumour incidence of Han:SPRD rats. Z Versuchstierk 33: 23-28

Kaspareit-Rittinghausen J, Wiese K, Deerberg F, Nitsche B (1990b) Incidence and morphology of spontaneous thyroid tumours in different strains of rats. J Comp Path 102: 421-432

Löhrke H, Hesse B, Goerttler K (1982) Spontaneous tumours in male and female specified pathogen-free Sprague-Dawley rats (outbred stock Sut:SDT). Z Versuchstierk 24: 225-230

Mohr U, Bader R, Ernst H, Ettlin R, Gembardt C, Harleman JH, Hartig F, Jahn W, Kaliner G, Karbe E, Kaufmann W, Krieg K, Krinke G, Küttler K, Landes C, Mettler F, Morawietz G, Notman J, Püschner H, Qureshi S, Reznik G, Rittinghausen S, Tuch K, Urwyler H, Weisse G, Weisse I, Zehnder J (1990) Tumor Registry Data Base: Suggestions for a systematized nomenclature for pre-neoplastic and neoplastic lesions in rats. Exp Pathol 38: 1-18

Morawietz G, Rittinghausen S, Mohr U (1990) Tumor Registry Data Base: Advantages using a systematized nomenclature. Schweiz Arch Tierheilk 132: 448-449

Morawietz G, Rittinghausen S (1990) Tumor Registry Data Base: Applications in modern toxicological pathology. Presented at the 10th ILSI Histopathology Seminar on the Musculoskeletal System of Laboratory Animals, Hannover, Germany, Sept 5-7, 1990

Suzuki H, Mohr U, Kimmerle G (1979) Spontaneous endocrine tumors in Sprague-Dawley rats. J Cancer Res Clin Oncol 95: 187-196

Turusov VS, Mohr U (eds) (1990) Pathology of tumours in laboratory animals. Vol I. Tumours of the rat, 2nd edition. IARC Scientific Publications No. 99, Lyon

Modelling

(Chair: H. G. NEUMANN, Würzburg)

BIOLOGICALLY BASED MODELING IN TOXICOLOGY RESEARCH

M E Andersen, K Krishnan, R B Conolly and R O McClellan

Chemical Industry Institute of Toxicology
Research Triangle Park, NC 27709

ABSTRACT

Biologically based modeling can be described as the process by which the specific mechanistic steps governing tissue disposition and toxic action of chemicals are expressed in quantitative terms by a set of equations leading to prediction of the outcome of specific toxicological experiments by computer simulation. These models are useful in risk assessment because their mechanistic biological basis permits the high-to-low dose, route to route and interspecies extrapolation of the tissue disposition and toxic action of chemicals. By far their greatest utility is not as "finished" risk assessment models, but as research tools that convey a quantitative expression of our hypotheses of tissue disposition and toxic action of a chemical. A structured modeling approach to toxicology problems helps identify the data gaps in the areas of chemical disposition and toxic action, thus prioritizing on-going research to obtain critical information required to conduct quantitative risk assessment. This paper examines progress in developing comprehensive biologically based models for cancer induction by non-genotoxic carcinogens that are cytotoxic in target tissues. The strategies for linking the models on dosimetry, cytotoxicity, and carcinogenicity are described in detail. The basic concepts and approaches discussed here can be applied to many other toxic chemicals and to toxicity endpoints other than cancer.

INTRODUCTION

The objective of toxicology research is to infer risk to exposed people based on observations of toxic effects in test systems. This goal is most readily achieved by interdisciplinary research programs examining the mechanistic steps in the various aspects of the exposure - dose - response continuum for a particular substance, with emphasis on extrapolation, i.e., on the manner in which critical determinants of the biological processes vary among species. The efforts in the individual disciplines need to be closely co-ordinated to insure a human health risk assessment focus in the research. Integration of individual activities in toxicological research programs can be significantly enhanced by the development of explicit, biologically based (BB) models of the toxicological phenomena under investigation. These BB models essentially become quantitative, testable hypotheses about the mechanisms of toxicity and expressly convey the proposed interrelationships of the individual research activities in the overall research program.

Comprehensive BB models for toxic effects will generally include descriptions of three processes: tissue dosimetry, i.e., the delivery of toxic chemical to target tissues; biochemical mechanism of action, i.e., the nature of the interaction of the toxic chemical with cellular materials; and, the toxic response itself, i.e., the events that link the cellular changes with the appearance of overt deleterious effects in the organism (Conolly et al. 1988). This paper emphasizes the sequential linkage of these three types of models in providing a complete computer model for cancer induction by cytotoxic chemicals, such as chloroform or carbon tetrachloride, and discusses the use of these models in providing a risk assessment orientation to the various research activities eventually needed to confirm the accuracy of the model structure.

BIOLOGICALLY-BASED TISSUE DOSIMETRY MODELS

Biologically based dosimetry models are integrated mathematical descriptions of the complex interplay of physiological and chemical-specific factors involved in the absorption, metabolism, distribution and elimination of chemicals. In the past twenty years, there has been steady progress in providing increasing biological realism to the pharmacokinetic models to predict tissue dose of the active forms of chemicals (Bischoff and Brown 1966; Bischoff et al. 1971; Ramsey and Andersen 1984; Andersen et al. 1987). These models generally are referred to as physiologically-based pharmacokinetic (PB-PK) models. Early work with PB-PK models was conducted by chemical engineers examining the distribution of anti-neoplastic drugs (Himmelstein and Lutz 1979; Gerlowski and Jain 1983). More recently, these models have found broad applications in toxicology research (Leung 1991) and in risk assessment for chemicals of environmental and occupational interest (Clewell and Andersen 1985). PB-PK models for volatile chemicals such as chloroform, consist of a series of tissue compartments each receiving the chemical in arterial blood and losing it via venous effluent. These biologically relevant compartments, arranged in an anatomically accurate manner, are defined with appropriate physiological characteristics. The compartments may represent a single tissue, or a tissue group which consists of several separate tissues with similar blood flow and solubility characteristics.

In the PB-PK models for low molecular weight volatile substances, tissue uptake is described as a flow-limited process (Andersen 1991). The rate of change in the amount of the chemical in each tissue compartment is described with mass balance differential equations of the following form:

$$dA_i/dt = V_i dC_i/dt = Q_i(C_{art} - C_{vi}) = Q_i(C_{art} - C_i/P_i) \qquad [1]$$

where

dA_i/dt = Rate of change in the amount of a chemical in the tissue i (mg/hr)

V_i = Volume of tissue (L)

Q_i = Blood flow to the tissue (L/hr)

C_{art} = Concentration of the chemical in arterial blood (mg/L)

C_{vi} = Concentration of the chemical in the venous blood leaving the tissue (mg/L)

C_i = Concentration of the chemical in the tissue (mg/L)
P_i = Tissue:blood partition coefficient

Tissue metabolism is included by introducing one or more loss terms for first-order, second-order or saturable reactions, as appropriate. For instance,

$$dA_i/dt = Q_i(C_{art} - C_{vi}) - dA_{met}/dt$$
$$dA_{met}/dt = V_{max}C_{vi}/(K_m+C_{vi}) + K_fC_{vi}V_i \quad [2]$$

where
dA_{met}/dt = Rate of amount of chemical metabolized (mg/hr)
V_{max} = Maximum enzymatic reaction rate (mg/hr)
K_m = Michaelis constant for enzymatic reaction (mg/L)
K_f = First order rate constant for metabolism (hr^{-1})

Three types of information are required for these PB-PK models. (1) Physiological parameters - such as alveolar ventilation rate, cardiac output, blood flow rates and tissue volumes - can be obtained from biomedical literature (Arms and Travis 1988). (2) Partition coefficients, indicating the relative ratio of concentration between two media (e.g., blood and air, or tissue and blood) at equilibrium, can be estimated for volatile chemicals in vitro by vial-equilibration methods (Sato and Nakajima 1979; Gargas et al. 1989). (3) Biochemical parameters required for these models are mainly the kinetic constants for biotransformation, macromolecular binding and elimination processes.

The rate constants for in vivo metabolism of volatile chemicals can be determined by kinetic studies in closed-atmosphere gas uptake studies (Filser and Bolt 1979; Andersen et al. 1980; Gargas et al. 1986, 1990). Kinetic biotransformation parameters can also be approximated by in vitro experiments with cells, tissue homogenates, or sub-cellular fractions (Reitz et al. 1988). These in vitro approaches, though, can be misleading because physiological factors, including blood flow or breathing rate, may be rate-limiting for chemical metabolism in vivo (Wilkinson and Shand 1975; Andersen 1981). Kinetic constants collected with in vitro preparations need to be carefully interpreted in light of the integration of these parameters within the context of an intact animal i.e., within a health risk assessment framework.

Once formulated, by combining information on animal physiology, kinetic constants for biotransformation reactions, and partition coefficients, the PB-PK model can simulate the pharmacokinetics of the chemical in the test species by numerical integration with a digital computer. Computer predictions are compared with actual experimental data from the animals to test the consistency of the model (Clewell and Andersen 1986).

The major advantage of PB-PK models over the traditional pharmacokinetic models is their extrapolative ability. Validated PB-PK models can be used to predict the behavior of chemicals administered by varying exposure routes, at differing doses, and even in different species. Exposure route extrapolations are performed by adding appropriate input terms to the set of differential equations. High dose to low dose nonlinearities are often accounted for by the saturable kinetics of biotransformation, excretion, or protein binding. Other dose-dependencies, associated with enzyme induction, enzyme inactivation, depletion of glutathione reserves, and the overt toxic effects at high doses have all been described with PB-PK models (Krishnan and Andersen 1991).

The mechanistic and biological realism of these pharmacokinetic models also provide the basis for conducting interspecies extrapolation. With these PB-PK models interspecies scaling is done by substituting the rodent physiological parameters with human values and by determining the appropriate chemical-specific, and biochemical parameters for humans (Dedrick 1973; Reitz et al. 1988).

BIOLOGICALLY BASED MECHANISMS OF ACTION MODELS - CYTOTOXICITY

Biologically based models for the mechanisms of toxic action involve mathematical descriptions of the quantitative interrelationship between critical elements of the series of events leading to adverse tissue response. Such descriptions have been developed for chloroform, a frequent contaminant in water supplies where chlorine is used for disinfection. Mice dosed with chloroform in corn oil develop liver toxicity and liver tumors (National Cancer Institute 1976). In the liver, this chemical is oxidatively metabolized to phosgene, a highly reactive intermediate, releasing a total of three moles of hydrochloric acid per mole chloroform. Corley et al. (1990) developed a PB-PK model for chloroform and examined the correlation between toxicity and tissue dose with respect to metabolized chloroform.

Hepatotoxicity correlated with the rate of metabolism of chloroform and not, for instance, with the amount of chloroform metabolite bound to tissue macromolecules (Reitz et al. 1990). This observation suggests that $CHCl_3$ toxicity is due to transient concentrations of acid and highly reactive metabolites, and not to persistent cellular adducts. The determinants of tissue dose then would be the instantaneous rate of metabolism (or the amount metabolized over a relatively short time period) and the volume of tissue in which the reactive metabolites are found.

The biochemical mechanisms model for cytotoxicity proposed by Reitz et al. (1990) links the likelihood of cell death with the rate of metabolism via a statistical distribution of cell sensitivities. Specifically, a normal distribution was used to calculate the proportion of cells susceptible to dying at any given rate of chloroform metabolism. In general terms, the linking relationship for the proposed biochemical mechanism by which liver cells are killed by chloroform is:

$$dN_h/dt = - SENS\ N_h\ k(death) \qquad [3]$$

where

SENS = Normal or log normal distribution of cell sensitivity to cytotoxicity as a function of $(dA_{met}/dt)/V_l$. This provides an estimate of the proportion of cells at risk at any time.

N_h = Total number of viable hepatocytes

k(death) = Rate at which cells at risk die

V_l = Liver volume

The extrapolation to people then requires quantitative information on the rates of metabolism of chloroform to phosgene and the liver volume. The rate is determined by V_{max}, K_m, liver chloroform concentration, and blood flow. Information on the kinetic constants of chloroform oxidation in people versus mice, together with the PB-PK model, permits prediction of the exposure conditions under which people will develop liver toxicity from this chemical. Because oxidation rates of these volatile xenobiotics are greater in mice than in people (Andersen et al. 1987; Reitz et al. 1990), people are expected to be at less risk than mice for hepatotoxic effects of chloroform. This conclusion is not necessarily expected to be valid with cytotoxic chemicals with other biochemical mechanisms of hepatotoxic action.

Another example, of a biologically based mathematical description of the mechanism of action, is the one developed for carbon tetrachloride. Reactive metabolites of this chemical bind to cellular macromolecules, and the accumulation of alkylated macromolecular adducts (with attendant decrease in concentration of functional macromolecules) is believed to initiate events leading to cell death. In this case, the level of chemically-stable adducts is related to both the rate of carbon tetrachloride metabolism and the degradation rates of the altered macromolecules. In contrast to chloroform, the relative adduct load over a more prolonged time period will have to be used as tissue dose in the distribution of cell sensitivity (Equation 3). Differential equations used in this case to describe the change in adduct levels with time will be of the following form:

$$dAdduct/dt = P_1 \, dA_{met}/dt - k_1 \, Adduct \qquad [4]$$

where

P_1 = Proportion of metabolite binding covalently

k_1 = Rate constant for adduct degradation or repair

In this case, the sensitivity to cell death would become a normal or log normal distribution based on adduct level in the cell. Relative degradation rates of macromolecular adduct in different species will also be important for interspecies comparisons. Paustenbach et al. (1988) developed a PB-PK model for carbon tetrachloride that included adduct compartments for metabolites and, indirectly, a degradation rate of the adducts. The turnover of cis-platin protein adducts has been explicitly included in a physiological model for this chemotherapeutic drug (Farris et al. 1988). For both chemicals, adduot degradation half-lives are close to one day in the rat. Information on the degradation rate constants of these adducts is lacking in people.

Thus, we see that for each chemical the biological determinants of toxicity can be sketched in a general way and used as a basis of model generation to articulate the nature of the unresolved issues. For instance, is the proposed relationship for cell death rate biologically plausible? How well does the normal distribution of cell sensitivities based on rates of metabolism reflect actual toxic responses seen when cells are treated with chloroform in vitro? Are there better dose metrics than the ones modeled? Most importantly, what experiments can be done to provide quantitative answers to these questions, to increase our confidence in our mechanistic hypothesis, and point the way toward developing strategies for interspecies extrapolation of the observations?

BIOLOGICALLY-BASED CANCER MODELS

With chloroform and carbon tetrachloride, a critical toxic effect of concern for human risk assessment is cancer. The role of cytotoxicity and resulting cell proliferation as a principal event in cancer induction is not easily incorporated into the conventional linearized-multistage treatment of the cancer dose response curves as used in the past by the US Environmental Protection Agency (Butterworth 1989). This biological information can be utilized in BB cancer models that specify exact roles for DNA alterations and cell replication in the production of tumors.

The two stage, cell growth model for cancer (also commonly called the Moolgavkar-Venzon-Knudson (MVK) model) specifies that two mutational events are needed to produce cancer - one converting a normal cell (N_0) to an intermediate cell phenotype (N_1), and a second mutation converting the intermediate cell to a fully malignant phenotype (N_2). The MVK formulation also accounts for the growth of normal and intermediate cell populations over time (Moolgavkar and Venzon 1979; Moolgavkar and Knudson 1981). This model adequately describes the incidence of certain human cancers, such as retinoblastoma, where the cancer is manifest after the loss of function of both allelles of a tumor suppressor gene in the malignant cell. Much more research is necessary to determine if this model is an adequate description of carcinogenic effects due to chemical carcinogens in animals, but at the very least it provides a testable hypothesis that permits introduction of biological data into dose-response analysis. In the MVK model as implemented by Conolly et al. (1988), mutation rates (mutations/division) per locus are expressed as the product of a mutation frequency (μ - probability of mutation/ division) and the replication rate (α - divisions/ time).

This completed model implementation for chloroform carcinogenesis then couples the dosimetry model which estimates rates of chloroform metabolism for different exposure conditions with a model for the mechanism of cell death related to the cellular burden of phosgene and HCl.

The output of this quantitative, mechanistic model for cell death (Equation 3) is then used in the MVK model to describe cell death and birth rates associated with repetitive chloroform exposure. (If the damage to the liver is not severe, it is assumed in the model that there is subsequent cell replication to restore the injured liver.) The birth rate of these cells becomes one of the inputs into the tissue response model for cancer. A deterministic equation approximating the rate of appearance of malignant cells (Conolly et al. 1992) is then used for calculating tumor dose-response curves:

$$dN_2/dt = \mu_2\, \alpha_2\, N_1[N_2{=}0] - \mu_2\, (N_1[N_2{=}0])^2) \int \alpha_2 dt / \int (\mu_1\, \alpha_1\, N_1) dt \qquad [5]$$

where

μ_2	=	Mutation probability for intermediate cell
α_2	=	Birth rate for intermediate cell
$N_1[N_2{=}0]$	=	Expected number of intermediate cells given that there are no cancer cells in the tissue, also referred to as the conditional probability

The linkage from the mechanisms model to the cancer model is via α_2 and α_1. The hypothesis would be that the birth rates (α_1 and α_2) increase to compensate for the death of cells by cytotoxicity estimated by equation 3, and that chloroform has no direct mutational effect (no effect on μ_1 or μ_2). The death rates for normal and intermediate cells, β_1 and β_2, appear directly in the differential equations for the growth of normal and intermediate cells, but these equations have not been included here. This equation completes the development of a quantitative, biologically based computer model for cancer induction secondary to cytotoxicity in the liver. The hypothesis now is captured in a set of linked equations that calculate tumor outcome for a wide variety of exposure conditions in different species.

A question that comes immediately to mind is whether the model is correct. To some extent that really is not the relevant question at this point. The appropriate questions to ask are: Does the model formulation accurately reflect the proposed hypothesis for the mechanism(s) of action of chloroform? Does it accurately describe the quantitative interconnections of the various models for dosimetry, toxicity and carcinogenicity? Does it suggest appropriate experiments to confirm the proposed hypotheses? And, does it adequately provide an integrated framework to show the importance of individual research efforts in assessing the expected human health consequences of exposure to the test chemical?

DISCUSSION

This paper details a model for chemical carcinogenesis with cytotoxic chemicals. The concepts discussed for model development are equally applicable for emphysematous changes with ozone, teratogenic responses to dioxin, renal toxicity of uranium, and virtually any other toxic response. In all cases, an argument will be made that there is not enough known to justify an elaborate model. In each case, however, the challenge is to distill the variety of observations into a working hypothesis and then give that hypothesis a mathematical structure. This forces the investigative team to give clear definition to their understanding of the steps in the exposure-dose-response sequence and of the quantitative interrelationships of these individual steps.

This proposed model-based research strategy is really nothing more than a general systems approach to solving complex interdisciplinary problems in toxicology. As noted by Kac (1969),

"The main role of models is not so much to explain and predict - though ultimately these are the main functions of science - as to polarize thinking and to pose sharp questions."

For the risk assessment orientation in toxicology research these sharp questions relate to how pieces of work give information of the linkages in the exposure-dose-response relationships, how they inform us of the manner in which observations in animals can be used to predict human risk, or how to design new studies to estimate the likely sensitivity of human tissues.

These quantitative models allow us to test hypotheses by simulation. We can ask questions of the if ..then.. variety. If the model structure is correct and a reaction rate is varied, then what impact will that have on toxicity outcome? The model can be used then to generate quantitative predictions of the expected outcome. If some other result is observed in actual experimentation, the model has to be modified. Just as our ideas can change with new experimental observations/data, these biological models are dynamic constructs which are also continually updated by new information and then serve in another iteration as guides for designing new research studies.

There has been a traditional tension between model building activities and mechanistic research. Are n't these two functions separate? Mechanistic research independently provides the grist from which to generate the model assumptions. This is one approach, but a particularly inefficient one. The goal is to use the model (i.e., quantitative hypothesis) to assist in selecting and designing those mechanistic studies that are most likely to answer quantitative issues relevant to risk assessment. The interaction of modeling and experimental design can serve to prioritize research opportunities, streamline the research program, and maintain the focus of experts on the larger picture of human health relevance.

Improvement in computational resources in recent years has spurred development of tissue dosimetry and tissue response modeling in toxicology. Great strides have been made with PB-PK models; the field of response modeling has not moved along as quickly (Conolly and Andersen 1991). This is partially due to the more complicated nature of the biological events involved in responses, and to the persistent notion that the models can only be created once all the data are available and there is widespread agreement on the mechanisms of the responses. Response models need to be created early on in research programs. It's our firm belief that the greatest challenge for toxicologist involved in studies to project human health risks of chemicals is to clearly recognize the need to create quantitative strategies to show how the various steps in exposure-dose-response link together. The integration of the individual pieces of information (i.e., in vitro metabolism, pharmacokinetics, macromolecular interaction, oncogene activation, etc.) within a complete "exposure - dose - response" model provides the link between downstream and upstream elements of the paradigm and imparts a risk assessment orientation to the individual toxicology research efforts in interdisciplinary research programs.

REFERENCES

Andersen ME (1981) Saturable metabolism and its relationship to toxicity. CRC Crit Rev Toxicol 9: 105-150.

Andersen ME (1991) Physiological modeling of organic compounds. Ann Occup Hyg 35: 309-321.

Andersen ME, Clewell HJ, Gargas ML, Smith FA, Reitz RH (1987) Physiologically based pharmacokinetics and the risk assessment process for methylene chloride. Toxicol Appl Pharmacol 87: 185-205.

Andersen ME, Gargas ML, Jones RA, Jenkins LJ (1980) Determination of the kinetic constants for metabolism of inhaled toxicants in vivo using gas uptake measurements. Toxicol Appl Pharmacol 54: 100-116.

Arms AD, Travis CC (1988) Reference physiological parameters in pharmacokinetic modeling. Office of Health and Environmental Assessment, U.S. Environmental Protection Agency, Washington, D.C., NTIS PB 88-196019.

Bischoff KB, Brown RH (1966) Drug distribution in mammals. Chem Eng Prog Sym Series 62: 33-45.

Bischoff KB, Dedrick RL, Zaharko DS, Longstrth JA (1971) Methotrexate pharmacokinetics. J Pharm Sci 60: 1128-1133.

Butterworth BE (1989) Nongenotoxic carcinogens in the regulatory environment. Regul Toxicol Pharmacol 9: 244-256.

Clewell HJ, Andersen ME (1985) Risk assessment extrapolations and physiological modeling. Toxicol Ind Health 1: 111-131.

Clewell HJ, Andersen ME (1986) A multiple dose- route physiological pharmacokinetic model for volatile organic chemicalsa using ACSL/PC. In: Cellier FD (ed) Languages for continuous system simulation. Society for Computer Simulation, San Diego, p 95.

Conolly RB, Andersen ME (1991) Biologically based pharmacodynamic models: tools for toxicological research and risk assessment. Ann Rev Pharmacol Toxicol 31: 503-523.

Conolly RB, Reitz RH, Clewell HJ, Andersen ME (1988) Pharmacokinetics, biochemical mechanism and mutation accumulation: A comprehensive model chemical carcinogenesis. Toxicol Lett 43: 189-200.

Conolly RB, Monticello TM, Morgan KT, Clewell HJ, Andersen ME (to be published) A biologically based risk assessment strategy for inhaled formaldehyde. Comments Toxicol •

Corley RA, Mendrela AL, Smith FA, Staats DA, Gargas ML, Conolly RB, Andersen ME, Reitz RH (1990) Development of a physiologically-based pharmacokinetic model for chloroform. Toxicol Appl Pharmacol 103: 512-527.

Dedrick RL (1973) Animal scale-up. J Pharmacokinetic Biopharm 1: 435-461.

Farris FF, Dedrick RL, King FG (1988) Cis-platin pharmacokinetics: applications of a physiological model. Toxicol Lett 43: 117-137.

Filser JG, Bolt HM (1979) Pharmacokinetics of inhaled ethylenes in rats. Arch Toxicol 42: 279-292.

Gargas ML, Andersen ME, Clewell HJ (1986) A physiologically based simulation approach for determining metabolic constants from gas uptake data. Toxicol Appl Pharmacol 86: 341-352.

Gargas ML, Burgess RJ, Voisard DJ, Cason GH, Andersen ME (1989) Partition coefficients for low molecular weight volatile chemicals in various liquids and tissues. Toxicol Appl Pharmacol 98: 87-99.

Gargas ML, Clewell HJ, Andersen ME (1990) Gas uptake techniques and the rates of metabolism of chloromethanes, chloroethanes, and chloroethylenes in the rat. Inhalation Toxicol 2: 295-319.

Gerlowski LE, Jain RJ (1983) Physiologically based pharmacokinetic modeling: principles and applications. J Pham Sci 72: 1103-1126.

Himmelstein KJ, Lutz RJ (1979) A review of the application of physiologically based pharmacokinetic modeling. J Pharmacokinet Biopharm 7: 127-137.

Kac M (1969) Some mathematical models in science. Science 166: 695-699.

Krishnan K, Andersen ME (to be published) The role of physiological modeling in reducing animal use in toxicology research. In: Goldberg AM (ed) Alternative methods in toxicology. vol 8. In vitro methods and mechanisms. Mary Ann Liebert Inc., New York.

Leung HW (1991) Development and utilization of physiologically based pharmacokinetic models for toxicological research. J Toxicol Environ Health 32: 247-268.

Moolgavkar SH, Knudson AG (1981) Mutation and cancer. A model for human carcinogenesis. J Natl Cancer Inst 66: 1037-1052.

Moolgavkar SH, Venzon DJ (1979) Two event models for carcinogenesis: incidence curves for childhood and adult tumors. Math Biosci 47: 55-77.

National Cancer Institute (1976) Carcinogenesis bioassay of chloroform. NCI, Bethedsa, MD, National Technical Information Series No. PB 264018/AS.

Paustenbach DJ, Clewell HJ, Gargas ML, Andersen ME (1988) A physiologically based pharmacokinetic model for inhaled carbon tetrachloride. Toxicol Appl Pharmacol 96: 191-211.

Ramsey JC, Andersen ME (1984) A physiologically based description of the inhalation pharmacokinetics of styrene in rats and humans. Toxicol Appl Pharmacol 73: 159-175.

Reitz RH, Mandrela AL, Park CN, Andersen ME, Guengerich FP (1988) Incorporation of in vitro enzyme data into the physiologically based pharmacokinetic model for methylene chloride: implications for risk assessment. Toxicol Lett 43: 97-116.

Reitz RH, Mandrela AL, Corley RA, Quast JF, Gargas ML, Andersen ME, Staats DA, Conolly RB (1990) Estimating risk of liver cancer associated with human exposure to chloroform using physiologically based pharmacokinetic modeling. Toxicol Appl Pharmacol 105: 443-459.

Sato A, Nakajima T (1979) Partition coefficients of some aromatic hydrocarboins and ketones in water and oil. Br J Ind Med 36: 231-234.

Wilkinson GR, Shand DG (1975) A physiological approach to hepatic drug clearance. Clin Pharmacol Ther 18: 377-390.

QSAR STUDIES IN GENETIC TOXICOLOGY: CONGENERIC AND NON CONGENERIC CHEMICALS

Romualdo BENIGNI and Alessandro GIULIANI

Istituto Superiore di Sanita'
Laboratory of Comparative Toxicology and Ecotoxicology
Rome - Italy

The extrathermodynamic approach to structure-activity studies

The Quantitative Structure-Activity Relationship (QSAR) methods are an important tool for the study of biological action mechanisms and for the prediction of chemical activity. The QSAR studies, although originally developed at the end of last century, were spurred on in the Sixties by the contributions of Corwin Hansch, as well as other investigators. It has been found that, for congeneric chemicals, it is possible to quantitatively correlate biological activity with physical chemical properties, such as hydrophobicity, steric and electronic factors (Hansch, 1990). Congeneric chemicals are those agents that share a common skeleton, but have different substituents attached to it. This approach -called extrathermodynamic or the Hansch approach- has been subsequently applied to a large number of situations, and has been highly refined. A great advantage of this method is that the physical chemical properties of most chemicals can be theoretically estimated. In fact, these properties are additive in the sense that they can be considered the sum of the contributions of the various atoms and substructures composing the molecule: thus, they can be calculated using the knowledge of the chemical formula, and the contributions of the substructures - the contributions of most atoms and substructures are tabulated-. These characteristics have made this method quick and easily applicable (Martin, 1981).

The extrathermodynamic approach in genetic toxicology

The extrathermodynamic approach has been succesfully applied to many different problems; among them, there is also a large number of reports focusing on chemical mutagenicity and carcinogenicity (for reviews, see Frierson et al., 1986; Benigni et al., 1989a). The following are selected QSAR studies on mutagenicity, performed in Prof. Hansch's laboratory. These examples cogently demonstrate how QSAR analyses contribute to the elucidation of action mechanisms.

Lopez de Compadre et al. (1990) studied the mutagenicity of nitroaromatic compounds in Salmonella typhimurium. The resulting equation for TA100 strain mutagenicity was:

$$\text{Log TA100} = 1.36\ \text{LogP} - 1.98\ \text{E-LUMO} - 7.01$$

where TA100 indicates revertants/nmol of mutagen, LogP is a measure of the hydrophobicity of the molecule, and E-LUMO is the energy of the Lowest Unoccupied Molecular Orbital. Thus E-LUMO parametrizes the ability of the chemical to accept electrons. The LogP term mainly reflects the interaction of the chemicals with the lipophilic components of the cell, and/or the interaction with enzymes. Consequently, LogP is related to the ability of the chemical to be absorbed into and transported within cells and organs, and to the chemical availability for metabolic transformations. This term is found in QSAR's of many different phenomena: the presence of LogP in this equation indicates that the hydrophobicity (hence absorption and distribution) is a crucial requirement for the activity of nitroaromatic compounds in the TA100 Salmonella strain. The presence of the E-LUMO term in the equation is very interesting. Higher mutagenic activity is associated with lower values of E-LUMO, i.e. better electron acceptors. This is consistent with increasing ease of reduction. This is also consistent with the notion that the crucial step for mutagenic activity of aromatic nitro compounds is the metabolic reduction by one or more cytosolic nitroreductases.

Shusterman et al. (1990) studied the mutagenicity of triazenes, and found the following relationship:

$$\mathrm{Log}(1/C) = 0.95\ \mathrm{LogP} + 2.22\ \mathrm{E\text{-}HOMO} + 22.69$$

In this equation, C is the molar concentration of triazene that causes 30 mutations above background per 100 million TA92 bacteria, and E-HOMO is the energy of the Highest Occupied Molecular Orbital. E-HOMO parametrizes the chemical ability to lose electrons (e.g. in oxidative reactions). The form of the equation indicates that the more hydrophobic and the more electron rich triazenes are, the more active they are in the TA92 Ames test strain. These two factors are the same factors that favor hydroxylation by cytochrome P-450. This supports the hypothesis that the initial triazene activation by cytochrome P-450, and not the rate of DNA alkylation by R+ (activated compounds), governs the mutagenicity of these compounds.

The topological approach

In the same period in which Hansch elaborated his model, another method referring directly to the substituents -instead of to the physical chemical forces- was being formulated (Free and Wilson, 1964). This is the so-called Free-Wilson analysis, where the biological activity is directly expressed as a sum of the contributions of the various substituents (atoms and substructures attached to the basic molecular skeleton).

The Free-Wilson approach generated all the so-called topological approaches, which are based on topology, or 2-dimensional representation of the molecular structure. At present, several topological approaches exist: they differ from each other for the kind of molecular descriptors that they generate, and for the kinds of statistical mathematical methods used. The topological methods arise from 2 requirements: a) simplicity, because the molecular descriptors are obtained directly from the structural formula of the molecule. The generation of the descriptors is easily implemented in computer programs which can perform this operation automatically; b) the necessity to deal with non congeneric groups of chemicals, which share, however, the same kind of biological activity. Non congeneric chemicals are those chemicals which do not share a common skeleton, and belong to different chemical classes. The use of topological approaches with non congeneric chemicals is aimed at identifying some substructure or topological characteristic correlated with the action mechanism, and common to chemicals apparently very different from each other (Franke et al., 1985).

Topological approaches in genetic toxicology: examples

Among the different topological approaches applied in mutagenicity and carcinogenicity, it is worth considering ADAPT and CASE. They have developed a more precise configuration, and have been implemented in computer program packages that perform the various steps of the structure-activity analysis in sequence.

ADAPT (Automated Data Analysis by Pattern Recognition Techniques) (Stouch and Jurs, 1985) reads the structure formula of the molecule -which is coded in the computer- and generates a wide range of descriptors of three different classes: a) topological (number of the different types of atoms, bonds and rings; presence or absence of substructures; electronic charges on the substructures); b) geometrical, calculated on the basis of the 3-dimensional structure; c) whole-molecule, such as LogP. Once the descriptors of the molecule have been calculated, ADAPT applies different mathematical methods (pattern recognition) to indicate those correlated with the biological activity.

The CASE (Computer Automated Structure Evaluation) method (Klopman, 1984) has an underlying rational different than ADAPT and other topological approaches . The latter methods generate the molecular descriptors based on a library, or pre-defined list. CASE generates all the possible substructures consisting of 3 to 10 non-hydrogen atoms interconnected for each molecule. For a structure-activity study all the molecules are fragmented in this way, and the correlation of each substructure with the biological activity is screened. The substructures not correlated with the activity are eliminated, and a function as usually a linear one similar to the Hansch approach- including the relevant features is constructed. It is evident that the CASE method, which places no restrictions on the types of substructures generated, has the advantage of being able to identify important substructures never considered relevant before. A problem with this approach seems to be -at least in principle- the possibility of chance correlations. When analyzing a set of molecules with many atoms, the number of combinations of interconnected atoms increases exponentially: thus, the number of substructures screened may become very large, with a consequent increase in the probability of chance correlations (Topliss and Edwards, 1979). It is important in such cases to plan the study in such a way as to eliminate this danger. For example, with a data base of 1000 molecules no more than 200 substructures should be generated for the screening.

Extrathermodynamic versus topological approach

An important question concerns the difference, with relative advantages and disadvantages, between the approaches using physical chemical properties and topological descriptors. A first consideration is that the extra-thermodynamic approach, in its original formulation, cannot be used with non congeneric chemicals, while some topological methods can. Moreover, the topological methods have the advantage of simplicity in the generation of descriptors.

The advantage of the extra-thermodynamic approach, -when applicable- is in its ability to express solutions in terms of physical chemical forces, which are a consistent scientific theory. In this sense, these solutions are more general, and can be used for a clearer understanding of the activity mechanisms: thus, they also permit a better control of the validity and reasonability of the results of the analysis. On the contrary, the topological analysis produces empirical correlations between activity and presence of one or more substructures in the molecules. To extend the results of a topological analysis beyond the mere empirical correlation, it is necessary to interpret and "translate" them into more general concepts.

Further developments of QSAR approaches

Topological approaches derived from the fact that the classical extra-thermodynamic approach was not able to elucidate all structure-activity relationships. Other important developments followed, stimulated by the general advancement of science. Important methodological improvements were made in the field of statistical and mathematical treatment of the data. Other advances occurred on the side of description of structures and properties of molecules. Molecular Orbital calculations provided new sophisticated parameters. Molecular Mechanics methods, which were originally used for model building and conformational analysis of isolated and strictly gas phase molecules, were adapted to model intermolecular interactions, including both geometry and enthalpy of some model of drug-receptor interactions. In connection with Molecular Mechanics, computer graphics applications to QSAR have grown dramatically. Complex molecular interactions can now be investigated by building models on a graphics device, and optimizing the geometry with Molecular Orbital or Molecular Mechanics calculations (Tute, 1990).

The problem of non congeneric chemicals

The QSAR approaches were originally designed to study congeneric chemicals, but, soon attempts to apply them also to groups of non congeneric chemicals were performed. These attempts considered chemicals of different classes, inducing the same biological activity (for example, heterogeneous sets of chemicals inducing rodent carcinogenicity).

The difference between congeneric and non congeneric situations is fundamental. Congeneric chemicals are assumed to produce a certain biological action through the same mechanism; the different substituents of the molecules simply modulate the quantitative level of activity. Consequently, the QSAR approach is aimed at describing with an equation how these small variations of chemical structure -parametrized by physical chemical properties or topological indices- affect the biological activity levels (Hansch, 1977). The situation is basically different with agents of different chemical classes inducing the same kind of biological effect. Here, the chemicals are likely to induce the same effect (e.g. rodent carcinogenicity) through very different mechanisms, and it is extremely difficult -if not impossible- to summarize in one equation many different mechanisms, or to find some underlying principle common to all of them.

QSARs of non congeneric chemicals: examples

Inspite of such difficulties, many attempts for establishing QSAR's of non congeneric chemicals were made. The following is a presentation of a number of QSAR studies focusing on non congeneric chemicals: all these cases clearly exemplify the problems encountered in such types of applications.

Stouch and Jurs, 1985, studied a set of 105 compounds tested in E. coli WP2uvrA, which belonged to different chemical classes. A large number of descriptors were generated by ADAPT. A correlation was found between mutagenicity and 10 selected descriptors, including fragments, charges and geometric indices. The fit of the found QSAR equation to the original data was checked by performing a retrospective classification of the 105 compounds used for the construction of the correlation itself: 90% of the compounds were correctly classified as mutagens or non mutagens. Then the rule was applied to 40 external compounds, which had not been used for the analysis. Prediction results were poor: 50%, which is approximately what one would expect from random chance. It is evident that the descriptors and the QSAR found in the training set were not representative of the general underlying determinants of E. coli mutagenicity.

The above example focuses clearly on the crucial clue of the relationship between descriptive power and prediction in QSAR analysis. The ability to describe structure-activity relationships in a given sample of chemicals is a first requirement. But the ultimate objective and validation of a QSAR is the prediction of the activity of compounds not included in the training set (which is composed of the chemicals used to derive the QSAR), but still belonging to the same chemical context as the training set. A QSAR can be said to be really valid only when its predictions are succesfull, otherwise the QSAR is of no great usefulness or scientific meaningfulness.

QSAR's studies on non congeneric chemicals were also performed with the CASE approach. Rosenkranz and Klopman (1990) analyzed with CASE a portion of the US National Toxicology Program (NTP) Salmonella mutagenicity data (243 chemicals). CASE identified 13 structural determinants highly correlated with mutagenic activity. Based on these determinants, CASE correctly classified 99% of mutagens and 95% of non mutagens (retrospective classification of training set chemicals). The CASE approach has also been applied to a Gene-Tox derived Salmonella mutagenicity data base (808 chemicals) (Klopman et al., 1990): 29 activating and 3 inactivating structural determinants were identified. The retrospective classification based on these determinants was correct for 100% of mutagens and 70% of non mutagens. Recently, Dr. Zeiger (report at the 1991 Annual Meeting of the US Environmental Mutagenicity Society; manuscript in preparation) coordinated an exercise where four QSAR approaches were used to predict the Salmonella mutagenicity of 100 NTP chemicals that had not previously been published. Among others, also the CASE approach was included in the exercise. It was applied in two different versions: the first one using as a training set the previously published NTP chemicals (Rosenkranz and Klopmann, 1990), and the second one using as a training set the Gene-Tox data (Klopmann et al., 1990). In the first application, CASE correctly predicted the activity of 67% of mutagens and 84% of non mutagens. In the second application, CASE correctly identified 78% of mutagens and 66% of non mutagens. The gap between the performances on the training set (always very high, more than 90% accuracy) and the performance in the predictive experiment (much lower, about 70% accuracy) is evident.

Another example comes from our own experience with the structure-activity studies of non congeneric carcinogens. We have used whole-molecule physical chemical parameters: LogP, Molar Refractivity, Molecular Weight, G (free-energy of drug-receptor interaction, parametrizing the chemical's ability to bind with biological receptors, such as enzymes), and electrophilicity. The hypothesis that we wanted to test was whether active molecules and inactive molecules would show different combinations of parameter values: the optimal combination of physical chemical values for being a carcinogen would be different than that for being a non carcinogen. We have also devised a new mathematical approach (Relationship Pattern Discrimination, REPAD), a pattern recognition method specific for non linear situations. Of a set of 137 selected carcinogens and non carcinogens of different chemical classes, REPAD correctly classified 90% of the chemicals of the training set (retrospective classification) (Benigni et al., 1989b). After this internal validation, we tried to predict the activity of more than 100 chemicals whose experimental carcinogenicity was known (NTP experimentation), but had not been used by us in the derivation of the REPAD discriminant equations. The application of the QSAR rule found for the training set was only 60% correct in predicting the carcinogens and non carcinogens. No better prediction was achieved inspite of several attempts and modifications of the mathematical procedure (our unpublished results).

In the preceding examples, applications of 3 different approaches to the QSAR of non congeneric chemicals have been presented. One is entirely based on topological descriptors (CASE), the second one (ADAPT) is based on a combination of topological and physical chemical descriptors, the third one (REPAD) uses only physical chemical descriptors. In the 3 cases, a good QSAR for the training set has ben found (more than 90% accuracy as retrospective classification); however these QSARs, when used to predict the activity of further chemicals, have all shown much lower performances, thus indicating a lack of general validity.

There could be several reasons for this failure. For example, for the topological approaches (such as CASE) some specific reason may be hypothesized. A first consideration is that the CASE system identifies all possible substructures of a set of chemicals: this number is usually very high (for example, 808 chemicals generated 293713 different fragments (Klopman et al., 1990)), and may give rise to chance correlations (Topliss and Edwards, 1979). Second, an assumption underlying most topological methods is that each substructural unit always gives the same type of contribution to a given biological activity: on the contrary, the influence exerted by a substructure may be very different in different chemical environments (i.e. in different types, or classes of molecules) (Franke et al., 1985). Together with specific reasons, such as those exemplified above, at the origin of the lack of good predictive power there is the general problem of the difficulty of the definition of a QSAR simultaneously valid for many different chemical classes and biological mechanisms. (Obviously, these considerations only reflect the present state of the art, and do not preclude future developments in the field of QSAR of non congeneric chemicals).

QSARs for non congeneric chemicals: perspectives

In our opinion, the difficulties and problems outlined in the preceding discussion indicate that QSAR studies for non congeneric chemicals should be viewed in a basically different way than QSAR for congeneric chemicals. With congeneric chemicals, a continous variation of chemical properties determines a continuous variation of a biological activity dictated by one mechanism of action; in this case, a precise QSAR description of the phenomenon should be expected. For non congeneric chemicals, there is not such a simple underlying pattern; as a consequence, the QSAR study is expected to produce a less defined and precise result. In other words, what we can expect is a probabilistic estimate of the type of biological action, rather than an accurate prediction of the activity level.

However, non congeneric QSAR approaches have their own role and importance, provided that their limitations are duly kept in mind. This point may be clarified with the following examples.

A first example is represented by the Ke approach. In the laboratory of Dr. Bakale, the relative electrophilicity (Ke) of a series of carcinogens and non carcinogens has been experimentally measured with a Pulse-Conductivity technique (Bakale and McCreary, 1987; 1990). We have recently reviewed the Ke data published for 206 chemicals. If a Ke = 3.0 x 10^{12} $M^{-1}s^{-1}$ is taken as a threshold between electrophilic (high Ke) and putatively carcinogenic, and non electrophilic (low Ke) and putatively non carcinogenic chemicals, the Ke system correctly predicted the activity of 67% carcinogens and 79% non carcinogens (total accuracy 72%). Moreover, 78% of chemicals positive in the Ke (high Ke) system were carcinogens (positive predictivity). In other words, a chemical with high Ke has a high probability of being a carcinogen, whereas a chemical with low Ke has a high probability of being a non carcinogen (manuscript in preparation). We have further analyzed the data by dividing the Ke values of the 206 chemicals into four intervals. The chemicals in the class with lowest Ke values resulted to include a number of non carcinogens about two times higher than that of carcinogens. Chemicals with Ke values ranging in the second interval had about the same probability of being active or non active: this can be considered an area of uncertainty. In the third interval, there were two times more carcinogens than non carcinogens. The fourth interval of Ke values (highest Ke) was the most interesting; there were about 6 times more carcinogens than non carcinogens.

This evidence indicates very clearly the role and usefulness of Ke for risk assessment. The Ke value for a chemical gives a probability of carcinogenicity; this probability can be used to rank the chemical in a priority scale for more detailed studies, either theoretical or experimental. In particular, a Ke value in the highest range would indicate a very urgent need for further in depth investigations.

Another example -specific for the identification of Salmonella mutagens- is represented by the Structural Alert approach. Ashby (1985) has screened the molecular structures of a large number of carcinogens, and has identified a number of substructures that are frequentely associated with the activity. These are called Structural Alerts (SA). However, it has been found that the presence of SAs does not explain the activity of all kinds of carcinogens, but has a strong correlation with the Ames test results. The conclusion was that the SAs characterize the genotoxic carcinogens, whereas do not correlate with nongenotoxic carcinogenicity (Ashby et al., 1989). The SA approach was included in the exercise coordinated by Dr. Zeiger, and permitted the correct prediction of 80% of Salmonella mutagens (sensitivity), and 65% of non mutagens (specificity). Moreover, 65% of chemicals containing a SA were Salmonella mutagens (positive predictivity). Thus, a SA-positive chemical deserves a high rank on a priority scale for further studies.

It should be noted that the SA approach had a satisfactory performance for identifying Salmonella mutagens, whereas did not work equally well for identifying rodent carcinogens. In fact, on the set of 206 chemicals studied with the Ke approach, we have calculated that SA correctly predicted only 53% of carcinogens and 65% of non carcinogens (manuscript in preparation). On the contrary, the performance of Ke was relatively good for the carcinogenic activity, whereas it resulted to be unsatisfactory for Salmonella mutagenicity: it was reported to have 60% accuracy in the exercise coordinated by Dr. Zeiger.

In conclusion, the role of QSAR approaches in the study of non congeneric chemicals should consist of producing rapidly and at low cost the classification of the chemicals for risk assessment and priority setting. Instead, well-tuned descriptions of biological phenomena should be expected from QSAR studies of congeneric chemicals.

References

Ashby, J. (1985) Fundamental structural alerts to potential carcinogenicity or non-carcinogenicity, Environ. Mutagen., 7: 919-921.

Ashby, J., R.W. Tennant, E. Zeiger and S. Stasiewicz (1989) Classification according to chemical structure, mutagenicity to Salmonella and level of carcinogenicity of a further 42 chemicals tested for carcinogenicity by the U.S. National Toxicology Program, Mutat. Res., 223: 73-103.

Bakale G. and McCreary R.D. (1987) A physico-chemical screening test for chemical carcinogens: the Ke test, Carcinogenesis, 8: 253-264.

Bakale G. and McCreary R.D. (1990) Response of the Ke test to NCI/NTP-screened chemicals. I. Non-genotoxic carcinogens and genotoxic non-carcinogens, Carcinogenesis, 11: 1811-1818.

Benigni, R., C. Andreoli, and A. Giuliani (1989a) Quantitative structure-activity relationships: principles, and applications to mutagenicity and carcinogenicity. Mutat. Res., 221: 197- 216.

Benigni R., Andreoli C., and Giuliani A. (1989b) Structure-activity studies of chemical carcinogens: use of an electrophilic reactivity parameter in a new QSAR model, Carcinogenesis, 10: 55-61.

Franke R., S. Huebel, and W.J. Streich (1985) Sustructural QSAR approaches and topological pharmacophores, Environ. Health Persp., 61: 275-285.

Free, S.M., and J.W. Wilson (1964) A mathematical contribution to structure-activity studies. J. Med. Chem. 7: 395-399.

Frierson, M.R., G. Klopman, and H.S. Rosenkranz (1986) Structure-activity relationships (SAR's) among mutagens and carcinogens: a review, Environ. Mutagenesis, 8: 283-327.

Hansch C. (1977) On the predictive value of QSAR, in J.A. Keverling Buisman (Ed) Biological activity and chemical structure, Elsevier, Amsterdam.

Hansch C. (Ed) (1990) Comprehensive medicinal chemistry, Vol. 4, Quantitative Drug Design, Pergamon Press, Oxford.

Klopman G. (1984) Artificial intelligence approach to structure-activity studies: Computer automated structure evaluation of biological activity of organic molecules, J. Am. Chem. Soc., 106: 7315-7321.
Klopman G., M.R. Frierson, and H.S. Rosenkranz (1990) The structural basis of the mutagenicity of chemicals in Salmonella typhimurium: the Gene-Tox data base, Mutat. Res., 228: 1-50.
Lopez de Compadre R.L., A. Kumar Debnath, A.J. Shusterman, and C. Hanšch (1990) LUMO energies and hydrophobicity as determinants of mutagenicity by nitroaromatic compounds in Salmonella typhimurium. Environ. Mol. Mutagenesis, 15: 44-55.
Martin, Y.C. (1981) A practitioner's perspective of the role of quantitative structure- activity analysis in medicinal chemistry. J. Med. Chem., 24: 229-237.
Rosenkranz H.S. and G. Klopman (1990) The structural basis of the mutagenicity of chemicals in Salmonella typhimurium: the National Toxicology Program data base, Mutat. Res., 228: 51-80.
Shusterman A.J., A. Kumar Debnath, C. Hansch, G.W. Horn, F.R. Fronczek, A.C. Greene, and S.F. Watkins (1990) Mutagenicity of dimethyl heteroaromatic triazenes in the Ames test. The role of hydrophobicity and electronic effects. Mol. Pharmacol., december: 939-944.
Stouch T.R. and Jurs P.C. (1985) Computer-assisted studies of molecular structure and genotoxic activity by pattern recognition techniques, Environ. Health Perspect., 61: 329-343.
Topliss J.G. and R.P. Edwards (1979) Chance factors in studies of quantitative structure-activity relationships, J. Med. Chem., 22: 1238-1244.
Tute M.S. (1990) Hystory and objectives of quantitative drug design. in C.A. Ramsden (Ed) Comprehensive medicinal chemistry, Vol. 4, Quantitative drug design, Pergamon Press, Oxford.

QUANTITATIVE AND QUALITATIVE STRUCTURE ACTIVITY RELATIONSHIPS IN DEVELOPMENTAL TOXICITY

J E Ridings

SmithKline Beecham Pharmaceuticals, The Frythe, Welwyn, Herts, UK

INTRODUCTION

Quantitative and qualitative structure activity relationships (QSAR'S) have been used for many years in the design of new drugs but it is only in the last 15 years that they have been used as mechanistic tools and to provide early hazard warnings in Toxicology. These applications have become more complex progressing from acute investigations, often in non-mammalian species where treatment and response are closely linked, to chronic studies of carcinogenicity. The application of QSAR's to studies of developmental toxicity represents the most testing challenge to date. These studies differ from other investigations in that, in vivo, two interrelated experimental systems (the mother and the foetus) are involved. There are many endpoints and many factors contribute to the effects observed. Considering the requirements for a valid QSAR, according to the ECETOC report of 1986, "The biological activity under consideration involves a single, simple mechanism and a well-defined endpoint", can structure activity relationships be applied to such a complex area as teratology investigations?

EXPERIMENTAL METHODS

Whilst some recent QSAR investigations of teratology have used computer databases, the majority have been conducted by means of whole animal studies and/or in vitro techniques such as embryoculture and micromass culture. Experimental methods can be considered in terms of three basic components:

CHEMICAL DESCRIPTION	EXPERIMENTAL SYSTEM	ANALYSIS METHOD

Chemical Description

The chemical descriptors (examples shown in table 1) are each compound's fingerprints which might be expected to describe all aspects of its action. To increase the chance of establishing a successful QSAR, descriptors should be selected from several areas e.g. steric, hydrophobicity, electronic etc. One of the major difficulties of establishing a QSAR is the selection of the most relevant properties; hydrophobicity is the single most useful QSAR descriptor and should be the first choice especially in teratology studies, where it provides some indication of the potential for placental transfer.

Table 1
Examples of Chemical Descriptors

Physical	**Quantum Chemical**
Melting point	Molecular orbital indices
Vapour pressure	Electron density
Dissociation constant	Bond reactivity
Electronic	**Hydrophobicity**
Hammett constant	Partition coefficient
Ionization potential	Water solubility
Steric	**Structural**
Molar refractivity	Substructures
Molecular shape	Molecular connectivity
Molecular volume	Atom fragments

Experimental System

Experimental systems in developmental toxicity studies may vary by species, duration of treatment, route of administration, dosage, and compound formulation, all factors that can influence the resultant effect. In establishing any QSAR these experimental variables must be maintained constant. A restricted dosing period (window dosing) is often used to reduce endpoint variability, particularly malformations, and also to enable comparisons to be made with QSAR's made *in vitro* e.g. day 11 *post coitum*. Equimolar dosing is the most common way of trying to maintain comparability of dosage in a chemical series although the embryonic dose has rarely been considered.

Analysis Method

Analysis methods can be broadly divided into five categories:

- Analysis of multiple maternal and foetal endpoints using classic segment 2 parameters e.g. bodyweight, foetal/embryonic loss, abnormalities, followed by intuition.
- An extra-thermodynamic approach to individual endpoints: Hansch analysis.
- A sub-structural approach: Enslein(TOPKAT), Klopman(CASE).
- A rule-base: DEREK.
- Pattern recognition using unsupervised learning.

(The basis of the extra-thermodynamic and sub-structural approaches is regression and discriminant analysis.)

The first two categories can be applied to experimental QSAR's as demonstrated in the following two investigations:

Twelve structurally related dopaminergic agonists (Fig. 1) were administered orally at pharmacologically equivalent doses to groups of pre-mated rats (Baldwin and Ridings (1990)). Treatment was from days 6-15 post coitum with caesarian examinations on day 21 post coitum, day 0 post coitum being the day of sperm-positive smear. Five of the twelve compounds were teratogenic, producing a similar syndrome of malformations; substitution with a trifluoromethyl group in the 4-phenyl ring appeared to confer teratogenic potential to this series. However the intuitive approach does not account for the teratogen containing the naphthyl ring. Maternal toxicity did not appear to play a role in the aetiology of the observed malformations.

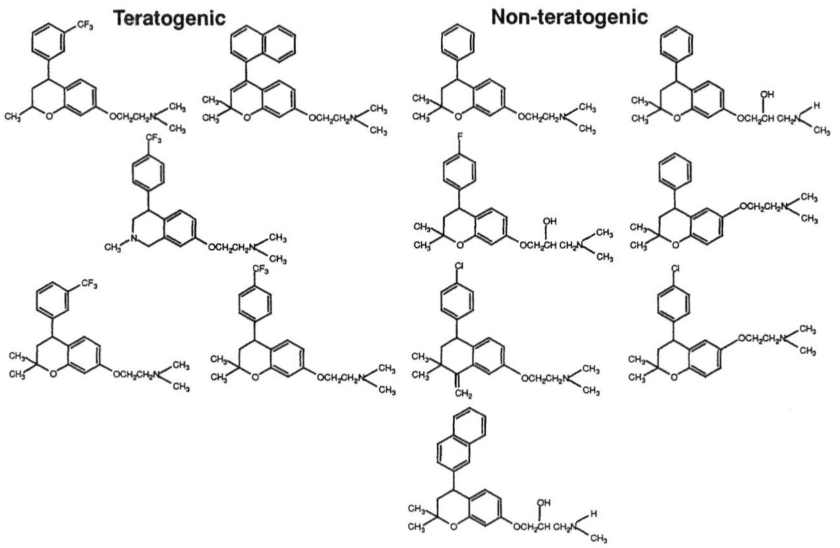

Twelve Structurally Related Dopaminergic Agonists
Figure 1

Twenty-two substituted para-phenols were administered orally to groups of pre-mated rats on day 11 post coitum, day 1 post coitum being the day of positive evidence of mating, using the same low, intermediate and high dose for each compound (Kavlock (1990). Maternal toxicity was assessed at 24hrs(MTOX1) and 72hrs(MTOX2) post-dosing. Peri-natal loss(PLOSS) was examined following parturition (Chernoff-Kavlock assay (Chernoff and Kavlock: 1987))

MTOX1 and MTOX2 were analysed relative to the chemical descriptors: hydrophobicity (Log P) and the Hammet constant(sigma) using Hansch analysis (Hansch and Leo (1979)).
The following equations were obtained:

$$1/MTOX1 = (0.0344 * \text{Log P}) + (-0.1503 * \text{sigma}) + 0.1195$$
$$n=22 \quad r=0.81 \quad p<0.001$$

$$1/MTOX2 = (0.0741 * \text{Log P}) + (-0.1858 * \text{sigma}) + 0.0589$$
$$n=22 \quad r=0.82 \quad p<0.0021$$

The MTOX1 model predicted that the potential to induce maternal toxicity increases with increasing lipid solubility(Log P) and decreasing electron-withdrawing properties(sigma) of the substituent on the para position of phenol. For MTOX2 the QSAR was similar in composition to that for MTOX1 indicating that the mechanisms resulting in toxicity at these time points is likely to be similar.

For PLOSS a QSAR could only be obtained for the eight most active compounds relative to Log P, sigma, and molar refractivity(MR) (MR was not included in the analysis of maternal toxicity as activity was unrelated to this parameter ($r=-0.21$ $p<0.338$)). PLOSS was again examined by Hansch analysis and the following equation was obtained:

$$1/PLOSS = (-0.2676 * \text{Log P}) + (-0.1827 * \text{sigma}) + (0.0265 * MR) + 0.4420$$
$$n=8 \quad r=0.82 \quad p<0.0021$$

The PLOSS model predicted that the decreased viability of implantation sites was related to decreasing lipid solubility, decreasing electron-withdrawing ability and increasing steric properties(MR) of the eight most potent para-phenols.

These three QSAR's indicated that, for the para-phenols, reduced viability was produced by a different mechanism than was maternal toxicity. No descriptor was identified in this experiment that distinguished the active set of compounds for PLOSS from the inactive set nor was any QSAR found for effects on litter weight.

This examination of the para-phenols demonstrated an advance from the qualitative evaluation of the dopaminergic agonists and showed the use of Hansch analysis not only to separately quantify maternal and foetal responses but also to indicate possible mechanisms of action.

Experimental methods, although providing familiar ground for most toxicologists, have a number of disadvantages in that they are time consuming and expensive, require the use of animals and costly amounts of compound and are often out of step with drug design. Computer generated QSAR's are routinely used in drug design and commercially available systems are being developed for the prediction of teratogenicity.

COMPUTER METHODS

In terms of the three components of the experimental methods (chemical description, experimental system and analysis method), computerisation of teratology analysis has involved substituting the experimental system with a database and replacing the Hansch method of analysis (Hansch and Leo (1979)) with a modified Free-Wilson method of analysis (Free and Wilson (1964)) based on substructural fragments. These methods can be considered in terms of three basic components:

CHEMICAL DESCRIPTION	EXPERIMENTAL SYSTEM	ANALYSIS METHOD

Chemical Description

The principles of chemical description remain. With the traditional Hansch approach, where there are pre-defined chemical descriptors, selecting the right descriptors can be problematical in examining a QSAR. While in one computer system (TOPKAT) there are some elements of this, there has generally been a move away from pre-defined descriptors.

Databases

Databases are the key to successful predictions of effect but for teratology studies they can at present be minefields for false predictions because:

- They suffer from limited and poor quality data - Enslein (1984) proposed that if data from the private files of pharmaceutical (and agrochemical) manufacturers were to become available for inclusion in models, the potential for improvement is considerable.
- Maternal toxicity is generally not taken into account.
- Many chemicals are inadequately described in terms of their endpoints of teratogenicity.
- The data within the database may be geographically biased.
- There is often a bias in the literature towards positive results.
- Databases are generally dysgeneric and subdivision to make them more congeneric may result in reduced predictivity.
- 'Grey' data may have varying interpretation.
- Updating of databases can be infrequent.

Kavlock (1988) summed up the database requirements stating "increasing amounts of data from in vitro and other short term assays have created the need for a peer-compiled and reviewed database of quantitative test results."

Analysis method

The Hansch method of analysis (Hansch and Leo (1979)) is dependent on the use of congeneric data sets and as computer databases often contain dysgeneric data an alternative method must be used.

The Free-Wilson approach (Free and Wilson (1964)) is based on the assumption that in a series of related chemicals each particular substructure adds a constant contribution to the biological activity of the parent molecule; perhaps a dangerous assumption in a system as complicated as the pregnant animal. Physico-chemical properties need not always be determined; they are contained within the sub-structural contributions. Enslein and Craig (1978) and Klopman (1984) have developed methods for predictions of teratogenicity based on the Free-Wilson approach (Free and Wilson (1964)) which are discussed below.

Commercially available systems for predictions of teratogenicity have been developed based on approaches by Enslein and Craig (TOPKAT) and Klopman (CASE).
Enslein and Craig: Enslein and Craig's approach has been applied to predictions of LD50, mutagenicity, carcinogenicity, teratogenicity and irritancy and a system based on this is being marketed by Health Designs Incorporated (HDi) as TOPKAT.

The teratogenicity model consists of four submodels: acyclic structures, heteroaromatic structures, alicyclic structures and carboaromatic structures. The model does not take into account maternal toxicity and the end point is teratogenicity in its broadest sense. It therefore produces a conservative estimate of the probability that a compound would demonstrate teratogenic effects were the assay performed.

The database contains information on 352 diverse structures assembled from Schardein's "Drug's As Teratogens" (1976), Shepard's "Catalog Of Teratogenic Agents" (1980) and from The Environmental Teratogens Information Centre (ETIC) collection. Data can be added to the base or it can be created de novo from 'in-house' data but the inherent equations and coefficients have to be re-computed. The database is subdivided as shown in Table 2.

Table 2
Number Of Compounds

Submodel	With Effects	Without Effects
Acyclics	26	43
Alicyclics	17	23
Carboaromatics	27	117
Heteroaromatics	22	77

(HDi information March 1989: latest data available)

Each compound is described by a set of structural and physical properties. These properties are then used in the construction of discriminant or regression equations which are used to estimate the probable teratogenic potential of any test compound.

Klopman: Klopman has developed the Computer Automated Structure Evaluation (CASE) method of analysis. The CASE system is not as yet commercially available although from a personal communication it is understood that a CASETOX program is in development containing databases from a number of sources including the Department of Environmental and Occupational Health of the University of Pittsburg; a limited human database will also be available. The databases can be expanded without affecting the subsequent analysis. Detailed information has not been released and, therefore, considerations of maternal toxicity and the proportions of active:inactive compounds cannot be made.

CASE differs from TOPKAT in that no pre-defined descriptors are required. Each molecule in the database is fragmented into all possible substructures of 3-10 linearly connected atoms. Fragments are then marked as active (e.g. teratogenic) or inactive (e.g. non-teratogenic) according to the activity of the parent molecule. All fragments (39 retinoids (Klopman and Dimayuga (1990)) generated 18,423 fragments, for example) are analysed to determine which of them have a disribution most skewed towards activity or inactivity. The ones with the greatest skew based on frequency occurrences are selected for the QSAR to yield a linear stepwise regression equation that best describes the activity of the compounds in the database. Klopman and Dimayuga (1990) claim to be able to distinguish cis and trans isomers.

CASE, by linking substructures to activity, offers the possibility of comparing substructures linked to maternal toxicity with those linked to embryotoxicity or teratogenicity. Frierson, Mielch and Kochhar (1990), analysed published data on a series of retinoids from a hamster teratogenicity assay and from a limb bud culture system and compared the substructural relationships between in vivo and in vitro effects. The substructural relationships were obtained in the two test systems were not the same suggesting that the mechanisms involved were different. This showed that CASE can be used not only as a predictor of teratogenicity but also as a possible mechanistic tool.

The strengths and weaknesses of both CASE and TOPKAT lie in their databases with the chance of false predictions being increased where the test compound is poorly represented. The degree of representation of the test compound/substructures must be considered when using a database model.

DEREK: A computer method called DEREK (Deductive Estimation of Risk from Existing Knowledge) is being developed by Sanderson and Earnshaw in collaboration with LHASA UK (1989) (the LHASA programme is the copyright of the President and Fellows of Harvard College and the University of Leeds) to allow the qualitative prediction of likely areas of possible toxicological risk for new or existing compounds, based on an analysis of their chemical structure. This is achieved by comparing the structure of the test molecule with a Rule Base of chemical structural features which are associated with particular toxicological properties. A warning is given for each of the features identified. The DEREK program does not provide a quantitative assessment of the risk, or a prediction that any potential risk identified will in fact occur. It only highlights areas in which toxicological problems are likely to arise, within the limitations imposed by the scope of the Rule Base. DEREK is an attempt to computerise "experience" or the "intuition" alluded to earlier by storing on a database scientific knowledge of structure activity relationships. There is very little application to developmental toxicology at present but a "Toxicology Working Party" to bring together data from individual pharmaceutical companies on all aspects of toxicology is in an embryonic form.

PATTERN RECOGNITION: The pattern recognition methods (unsupervised learning) of non-linear mapping and principal components analysis (Livingstone (1989)), routinely used by the drug design team at SmithKline Beecham, offer an alternative to the Free-Wilson and Hansch approaches. They provide qualitative correlations between a set of chemical descriptors and the presence or absence of a specific biological activity. Techniques are predictive and consideration of the chemical descriptors giving a good structure activity relationship may provide an indication of the mode of biological action. The pattern recognition approach would therefore appear to be ideally suited to teratology evaluations where the constant contribution of substructures may be speculative. There appear to be, however, no reports on the use of pattern recognition techniques in teratology evaluations but this approach is being examined in conjunction with our drug design team.

AN INDUSTRIAL VIEW OF QSAR

In order to determine the use of QSAR's for developmental toxicology assessments in the Pharmaceutical Industry, 13 companies were approached for their views.
The results of the survey are presented in Table 3.

Table 3
Survey of the use of QSAR in Developmental Toxicology
In 13 UK and European Pharmaceutical Companies

QSAR EVALUATED	(5)	**QSAR NOT EVALUATED**	(8)
Method		Reason	
In-house computer database	(1)	No Confidence	(1)
Literature	(3)	Never Considered	(4)
In vivo studies (Seg. 2)	(5)	Never Needed	(3)
In vitro studies	(3)		

()=Number of companies; more than one method may be described

This survey confirmed the lack of use of QSAR but two of the reasons given for not applying this method showed the void between the almost routine use of QSAR's in drug design and the application in reproductive toxicology:

> "We tried TOPKAT but it gave too many false results".
> "The mathematics involved is somewhat daunting for biological scientists".

These statements demonstrate the degree of ignorance that abounds in this area. A further statement in the 1990 FRAME Committee Report confirms this opinion:

> "So far the QSAR approach has not been successful in reproductive toxicology - reflecting lack of knowledge."

DISCUSSION

The answer to the introductory question, "Can structure activity relationships be successfully applied to such a complex area as teratology investigations ?", is undoubtedly "YES", provided that suitable criteria are applied, the results obtained are not taken as definitive and extrapolation across species is avoided. The principles and applications of QSAR's to mutagenicity and carcinogenicity described by Benigni, Andreoli and Giuliani (1989) can equally be applied to reproductive toxicology data. To use the single endpoint "teratogenicity" is acceptable as a first pass in determining potential hazard but Karnofsky's "Law" (Karnofsky(1965)) that everything is teratogenic under the right conditions must be kept in mind. To examine developmental toxicity as a single biological system, as done by TOPKAT and CASE, provides only preliminary information which should be followed up with more in-depth analysis. Teratology analysis can be viewed as toxicology in the mother for which a QSAR can be obtained and toxicology in the embryo/foetus for which there may be a separate QSAR. Comparison of these QSAR's may allow mechanisms of maternal and foetal toxicity to be separated, essential information in assessing potential teratogenic risk. Where structural abnormalities are produced each can be viewed as a separate endpoint for which, in theory, a QSAR can be obtained. The methods are available but the success of the QSAR is determined by how they are applied.

Experimental methods must be designed carefully to minimise the variability of the endpoints and where possible to consider the doses employed in terms not only of equipotency in the maternal compartment but also in terms of the foetal dose. Where known metabolites exist these should ideally be included in the series of compounds examined.

The databases of computer based predictive systems must be critically reviewed to ensure that they contain quality data in sufficient quantity to support QSAR analysis. For systems such as TOPKAT it would be preferable to expand the database extensively with quality-controlled in house data (both positive and negative); in its present form the database is too limited to provide reliable predictions of teratogenicity. Where insufficient relevant data are available in the database (even none) both TOPKAT and CASE make a prediction (CASE defaults to a positive prediction), with TOPKAT the user is made aware of the basis of "tentative" predictions. The DEREK database is an excellent attempt to centralise "experience" and should not be lightly dismissed. The only reservation in applying this approach to developmental toxicology is that it requires a rather advanced understanding of the relationships between individual chemical substructures and the endpoint. These relationships, in genetic toxicology and carcinogenicity, appear to be reproducible but for teratology evaluations the structural relationships are less concise.

The supervised learning methods of analysis - multiple regression and discriminant analysis are sound and may with care be used to relate most kinds of biological response, including teratology, to chemical structure. However, pattern recognition techniques provide an attractive alternative that should be considered in teratology analysis, where multiple endpoints are commonplace and the contribution of substructures to the resultant effect of the parent compound is less precise.

The complexity of mammalian development has and will continue to hinder the success of establishing QSAR's in this area but with careful consideration of the chemical descriptors, the experimental system/database and methods of analysis, good results could be obtained.

The application of QSAR's to the consideration of human risk assessment will undoubtedly occur in the future with the establishment of better databases of human exposure but how will a positive prediction in humans for teratogenicity be viewed where animal studies data are negative?
A quote from Leon Golberg (1983) equally applies to QSAR's in developmental toxicology:
""Double, double toil and trouble" surely continues ahead for toxicologists but we
should learn from Macbeth's experience and not take QSAR predictions too literally
even when they come from a computer rather than a witch."

REFERENCES

Baldwin J, Ridings J (1990) Structure and Teratogenic Activity Relationships Within A Series Of Dopaminergic Agonists. Teratology 42: 28A

Benigni R, Andreoli C, Giuliani A (1989) Quantitative Structure-Activity Relationships: Principles, and Applications To Mutagenicity and Carcinogenicity. Mutation Research 221: 197-216

Chernoff N, Kavlock RJ (1987) Further Evaluation Of An In Vitro Teratology Screen. Teratogenicity, Carcinogenicity and Mutagenicity 7: 7-16

Enslein K (1984) Estimation of Toxicological Endpoints by Structure Activity Relationships. Pharmacological Reviews 36: 131-135

Enslein K, Craig PN (1978) A Toxicity Estimation Model. J Environ. Pathol. Toxicol. 2: 115-121

European Chemical Industry Ecology and Toxicology Centre (1986) Structure Activity Relationships In Toxicology And Ecotoxicology: An Assessment. Monograph No. 8, Brussels, 24 February

Free SM, Wilson JW (1964) A Mathematical Contribution To Structure-Activity Studies. J. Med. Chem. 7: 395-399

Frierson MR, Mielch FA, Kochhar DM (1990) Computer-Automated Structure Evaluation (CASE) of Retinoids in Teratogenesis Bioassays. Fundamental and Applied Toxicology 14: 408-428

Golberg L (1983) Ed. Structure-Activity Correlation As A Predictive Tool In Toxicology. Hemisphere Publ. Corp., Washington

Hansch C, Leo A (1979) Substituent Constants For Correlation Analysis In Chemistry And Biology. Wiley, New York

Karnofsky D (1965) In, Wilson JG and Warkany J (Eds.) Teratology: Principles and Techniques, P185. Univ. Chicago Press, Chicago.

Kavlock RJ (1988) Structure Activity Relationships In Developmental Toxicology. Health Effects Research Laboratory Office of Research and Development, US EPA Publication, June

Kavlock RJ (1990) Structure-Activity Relationships in the Developmental Toxicity of Substituted Phenols: In Vivo Effects. Teratology 41: 43-59

Klopman G (1984) Artificial Intelligence Approach to Structure-Activity Studies. Computer Automated Structure Evaluation of Biological Activity of Organic Molecules. J. Am. Chem. Soc. 106: 7315-7321

References (cont.)

Klopman G, Dimayuga ML (1990) Computer Automated Structure Evaluation (CASE) of the Teratogenicity of Retinoids With The Aid of a Novel Geometric Index. Journal of Computer Aided Molecular Design 4: 117-130

Livingstone DJ (1989) Multivariate Quantitative Structure-Activity Relationship (QSAR) Methods Which May Be Applied To Pesticide Research. Pestic. Sci. 27: 287-304

Sanderson DM, Earnshaw CG (1991) Computer Prediction of Possible Toxicity From Chemical Structure; The DEREK System. Human and Experimental Toxicology 10: 261-273

Schardein JL (1976) Drugs As Teratogens. CRC Press, Cleveland, Ohio

Shepard TH (1980) Catalog Of Teratogenic Agents, 3rd Edition, John Hopkins Univ. Press, Baltimore

MULTI-COMPARTMENT MODEL TO STUDY THE EFFECT OF AIR-BLOOD AND BLOOD-TISSUE PARTITION COEFFICIENTS ON CONCENTRATION-TIME-EFFECT RELATIONSHIPS

A. Zwart, J.G.J. Lommen and V.J. Feron

TNO Toxicology and Nutrition Institute
P.O. Box 360, 3700 AJ Zeist, the Netherlands

ABSTRACT

The influence of the air-blood (λ) and blood-tissue partition coefficients of substances on the concentration-time-effect relationship $E=[aC^nT]^b$ was studied by model simulation of uptake and distribution of gases. The model consisted of a lung compartment and four tissue compartments. Rat and human model parameters were obtained from physiological data, and substance-dependent variables. The observable effect E (righting reflex, mortality, etc.) for directly acting systemic agents was supposed to be related to the (arbitrarily chosen) concentration of the substance in the arterial blood or in the liver compartment.
We conclude that for a directly acting systemic agent $n \geq 1$. If $n < 1$ other mechanisms such as metabolism must be incorporated in the model which, however, should not be excluded when $n \geq 1$. If the observable effect is related to arterial concentration of the substance, n increases with decreasing λ. If the liver concentration is related to the observable effect, man and rat may behave quite differently.
This has major consequences for the extrapolation of animal results to the human situation in risk assessment.

Keywords: modelling, extrapolation, risk assessment

INTRODUCTION

Awareness is growing that in evaluating labour safety requirements and hazard problems for the population after release of toxic substances, use should be made of concentration-time-effect relationships. Nevertheless, little insight has been gained so far into the factors determining these relationships. Such insight is of paramount importance for extrapolation of animal data to the human situation. Physiologically based pharmacokinetic (PBPK) models allow investigation of the determining factors of the concentration-time-effect relationships and the influence of specific differences to be investigated.
We studied the effects of air-blood, blood-tissue and differences in compartmental time constants on the concentration-time-effect relationships in a simple uptake and distribution model.

METHODS AND MATERIAL

Simulations were carried out with a general simulation program TUTSIM, developed by the University of Twente, the Netherlands. The model consisted of a lung compartment and four tissue compartments. The lung compartment was modelled as a co-current exchanger with a time constant zero, thus neglecting the uptake in the tissues of the lung. The four tissue compartments were modelled as described by Andersen et al. (1987). Time constants of these compartments depend on perfusion, tissue volume, and blood-tissue partition coefficients. We did not add metabolic activity to the liver compartment.
Simulations were carried out for styrene and methylene chloride with air-blood and blood-tissue partition coefficients according to Andersen et al. (1987) and for halothane with air-blood and blood-tissue partition coefficients according to Zwart et al. (1972). Concentration-time-effect diagrams for different substances and species were constructed by calculating, for various exposure concentrations, the time needed to reach a (arbitrarily chosen) concentration in arterial or liver compartments.

RESULTS

The concentration-time-effect diagrams for the rat model are shown in Fig. 1. Exposure times exceeding 0.2 h show no difference between liver or arterial concentration-related effects. The different points of concentration-time combinations to reach the effect form practically straight lines over the entire range of exposure times when using the arterial concentration as effect-related. For liver concentration-related effects a concavely curved line was obtained for exposure periods shorter than 0.2 h for halothane and styrene and shorter than 0.05 h for methylene chloride. The slopes of the curves increase with increasing air-blood partition coefficient.
The human uptake and distribution model was used in a slightly different way.
The time constants of the tissue compartments was taken to that of styrene. With this model we then looked for the effects of changing the air-blood

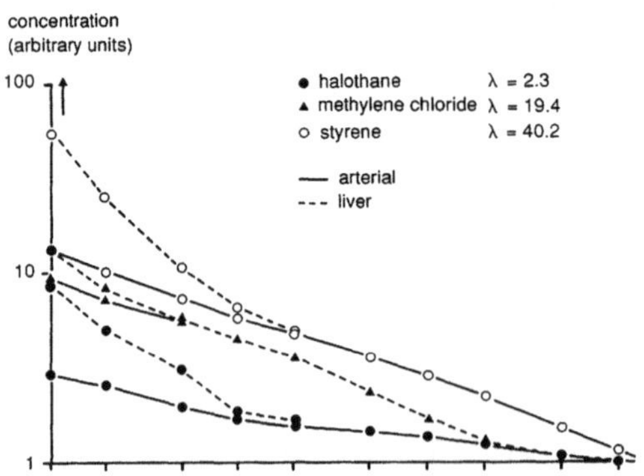

Fig. 1 Concentration-time combinations yielding a specific observable effect as simulated in a rat model

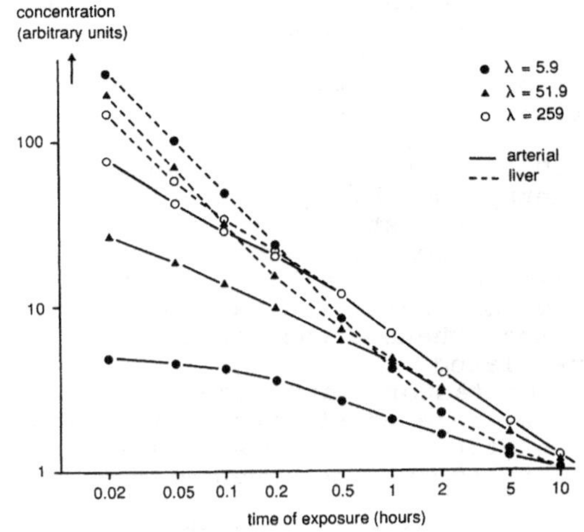

Fig. 2 Concentration-time combinations yielding a specific observable effect as simulated in a human model

partition coefficient. Results are shown in Fig. 2. The curves obtained with λ=51.9 correspond to the model values given by Andersen et al. (1987). In this human model the differences between the liver and arterial concentration-related effects are more pronounced than in the rat model. The slopes of the liver concentration-related curves are very close, and independent of the value of λ, whereas the arterial concentration-related curves show again an increase in the slope concomitantly with an increase in λ.

DISCUSSION

The concentration-time-effect relation is:

$$E=[aC^nT]^b \qquad \text{(Eq. 1)}$$

Eq. 1 is often described as the probit function:

$$P=ba + b\ln T + bn\ln C \qquad \text{(Eq. 2)}$$

When plotting combinations of ln$C$ a as function of ln$T$ for a given effect (e.g. LC50, P=5) a straight line is obtained with:

$$\ln C=(P-ba)-\ln T/n \qquad \text{(Eq. 3)}$$

Thus the slope of the curves as shown in Figs. 1 and 2 is inversely related to the factor n known from probit relations. In animal experiments concentration-time-effect diagrams often yield straight lines according to Eq. 3, but curves may also be shaped concavely or (rarely) convexly. Values of n may be larger than 1 but are also reported to be smaller than 1. Inspection of Fig. 1 shows almost linear curves for arterial concentration-related effects. Since n is inversely related to the slope, we can obtain a value of n from these slopes. Using the curve section between exposure times of 0.1 to 10 h of exposure we find the following values for n in the rat model: halothane λ=2.3, n=9.1; methylene chloride λ=19.4, n=3.0; styrene λ=40.2, n=2.8. Styrene in the human model (Fig. 2) yields λ=51.9, n=1.8. We see a decrease in n with increasing λ

which can partly explain the difference in n value for styrene in rat and man. The effect is, however, more pronounced since also the general increase of the time constants of the tissue compartments differs by a factor of five between rat and man. Short exposure times show a difference between liver and arterial concentration-related effects. This difference is present over a period of about 10 times the time constant of the liver compartment (T liver methylene chloride rat = 0.006 h, T liver styrene rat = 0.03 h, T liver styrene man = 0.16 h). As is clearly seen in Fig. 2, this factor is slightly dependent on λ. For extrapolation of rat data to human situations it is of great importance, particularly for gases that are poorly or moderately soluble in blood, to know whether the effect is arterial or liver concentration-related.
For λ=5.9 (halothane) in Fig. 2 the factor n increases from 1.2 for liver-related to 3.3 for artery-related effects whereas in the rat model in the normal range of exposure times (0.1 - 10 h) no difference is detected for halothane, not even with a lower λ.
The model used in this study did not incorporate metabolism (toxification or detoxification). Comparison of model simulations with *in vivo* obtained concentration-time-effect relationships allows to a certain extent the model to be falsified, showing that more complex models are needed to describe the *in vivo* results. The simple uptake and distribution model fails when $n < 1$ and for convex curvatures. This, of course, does not imply that the model is adequate when $n \geq 1$, or that in the case of a concave curve the substance reacts as a directly acting systemic agent with a liver concentration-related effect.

We conclude that PBPK modelling can be of help in interpreting *in vivo* obtained concentration-time-effect relationships and show which working mechanisms have to be known for a reliable extrapolation to the human situation. This is the more important since recently simple methods have been developed by Zwart et al. (1990) to determine concentration-time-effect relations which take about as many animals and costs as one conventional LC50 determination.

REFERENCES

Andersen ME, MacNaughton MG, Clewell III HJ, Paustenbach DJ (1987) Adjusting exposure limits for long and short exposure periods using a Physiological Pharmacokinetic Model. Am Ind Hyg Assoc J 48: 335-343

Zwart A, Smith NTy, Beneken JEW (1972) Multiple model approach to uptake and distribution of halothane: the use of an analog computer. Computers and Biomedical Research 5: 228-238

Zwart A, Arts JHE, Klokman-Houweling JM, Schoen ED (1990) Determination of concentration-time-mortality relationships to replace LC50 values. Inhalation Toxicology 2: 105-117

CONCEPTS FOR TOXICOKINETIC-TOXICODYNAMIC MODELLING IN CLINICAL TOXICOLOGY: APPLICATION TO ACUTE CARDIAC GLYCOSIDE INTOXICATIONS

M. URTIZBEREA*, A. SABOURAUD*, F. BAUD**, M. ROCHDI*, M. DEBRAY*** C. BISMUTH** and J.M. SCHERRMANN*

*INSERM U.26 Hôpital Fernand Widal, 200 rue du Faubourg Saint Denis 75010 PARIS (France)
**Clinique Toxicologique, Hôpital Fernand Widal
***Faculté de Pharmacie, 4 Avenue de l'observatoire, 75005 PARIS (France)

Relationships between pharmacokinetics (PK) and pharmacodynamics (PD) are now widely investigated for many drugs with the aim of improving the amount and the rate of drug delivery so as to control the duration and intensity of the pharmacological effect (Levy 1966; Holford 1981). Such relationships have not yet been investigated for toxic effects of drugs as occurs after acute self-poisoning. The establishment of relationships between kinetics of drugs at toxic doses (Toxicokinetics) and those of their pharmacological effects (Toxicodynamics) can be of interest in clinical toxicology. The aim would be to predict the occurrence and the intensity of toxic effects as a function of the ingested dose and the first plasma concentrations on admission of the patient to improve management. In this paper, we describe the different techniques of kinetic-effect modelling actually available, their applications in the case of acute digitalis poisoning and the limits of such techniques in clinical toxicology.

TECHNIQUES OF KINETIC-EFFECT MODELLING

Kinetic effect modelling have been widely described in reviews of Holford (1981)and Grevel (1990). The different events occurring after the administration of the drug may be summarized by the following scheme:

$$D \longrightarrow \begin{matrix} Ct(t) \\ Cp(t) \end{matrix} \longrightarrow Ce(t) \longrightarrow E(t)$$

The ingested dose of drug (D) produces plasma concentrations Cp(t) and tissular concentrations Ct(t) that can be described by pharmacokinetic models, effective concentrations at the action site Ce(t) which generate the effect E(t). If Cp(t) and E(t) occur simultaneously, it can be assumed that Ce(t) and Cp(t) are kinetically identical, i.e. the effect occurs in the central compartment. If a delay exists between maximum E(t) and Cp(t) observed, Cp(t) and Ce(t) are kinetically different. To explain such an observation, the existence of a compartment that is kinetically different from plasma and where the effect is produced can be postulated. Thus, from experimental plasma concentrations, concentrations in the peripheral compartment Ct(t) , i.e. tissular concentrations, may be mathematically calculated. In this case, it may be assumed that Ce(t) is equivalent to the tissue concentration of the peripheral compartment Ct(t), so that Ct(t) reaches a maximum at the same time as E(t). But there are other situations for which either a multi-compartment model is not adequate for modelling Cp(t) or a delay is still observed between the maxima of E(t) and Ct(t). In these cases, the effect compartment model proposed by Sheiner (1979) may be used. An effect compartment is linked to the classical compartment

models generated by pharmacokinetic studies. The action site compartment is linked to the central one by a first order input rate constant K1e and a first order output rate constant Keo. In the model of Sheiner, it is assumed that the effect compartment has a very small volume and that only very small amounts of drug would enter this compartment and would be excreted outside. It is also postulated that Ce equals Cp at steady state. In these conditions, it may be asserted that this compartment does not alter the kinetics of the other compartments.
When the effective concentration (Cp, Ct or Ce) is determined, its kinetics is related to E(t) using the different pharmacodynamic models commonly described :
- the linear model $E=SC$ or $E=SC+Eo$ where E is the effect, C the concentration of the drug, S a slope parameter and Eo the basal value of effect in the absence of drug.
- the log-linear model $E=SlogC + Eo$ where the log transformation effectively compresses the scale abscissa (Kelman 1980). This relationship is only available for effects ranging between 20 and 80% of maximal effect.
- the E_{max} model $E = E_{max}.C/EC_{50}+C$ where EC_{50} is the concentration leading to 50% of the maximal effect and E_{max} the maximum effect. This relationship takes into account the existence of a maximum effect.
- the sigmoid E_{max} model $E = E_{max}.Ce^n/EC^n{}_{50}+Ce^n$ where n is the slope parameter. This model enables to determine the steepness of the sigmoidal relationship.
PK-PD modelling can be performed with MKMODEL (Elsevier-BIOSOFT, Cambridge, UK) or SIPHAR (Simed, Créteil, France) softwares. The choice between the different models is based on statistical criteria such as Akaike and Schwartz criteria.

APPLICATION OF TOXICOKINETIC-TOXICODYNAMIC (TK-TD) MODELLING TO CARDIAC GLYCOSIDES

Pharmacological and toxicological effects of cardiac glycosides have been extensively investigated because of their low therapeutic index ratio. Binding of cardiac glycosides to Na^+K^+ ATPase is responsible for the inhibition of the active transport of Na^+ and K^+ which leads to pharmacological inotropic effects evaluated by measurement of haemodynamic changes (Shapiro 1970; Kramer 1978) and toxicological manifestations such as the increase of plasma K^+ and the decrease of red blood cell (RBC) K^+.
An easily detectable effect of the inhibition of Na^+K^+ATPase is the increase of plasma K^+ and the decrease of RBC K^+. Bismuth (1973) investigated the relationship between the plasma K^+ levels in the initial stages of acute digitalis intoxication and the digitalis plasma levels. They found a poor correlation (0.54) though a fatal prognosis was associated with hyperkalaemia (K^+ > 6mmol/l. The relationship linking plasma K^+ to plasma digitoxin concentrations in intoxicated patients was direct and TK-TD modelling is represented by a linear model (Urtizberea 1990). However, the plasma K^+ marker has two drawbacks. Firstly, when the delay between the ingestion of the toxin and the patient's arrival in intensive care unit is long, hyperkalaemia may not be noticed. Secondly, many therapeutic procedures aim at decreasing plasma K^+ and consequently prevent TK-TD modelling.
The depletion of RBC K^+ is a more convenient marker of digitalis intoxication because it occurs later (5 days after digitalis ingestion) and is always observed. In digitoxin-intoxicated patients, a direct correlation ($r=0.64$) between the calculated ingested dose of digitoxin and the maximum RBC K^+ depletion has been described (Urtizberea 1990). The main drawback of such a relationship is that bioavailability of the ingested dose in acute poisoning conditions is often difficult to appreciate. Loes (1978) demonstrated that digoxin therapy in children was associated with a significant decrease in RBC K^+. Similarly, Henion (1983) reported that the RBC ratio of Na^+ to K^+ in patients with digitalis intoxication was higher than in digitalis-treated patients. The main drawbacks of these studies is that they correlated single K^+ values with single cardiac glycoside concentrations; the pharmacokinetics of cardiac glycosides was not taken into account. The relationship between plasma digitoxin concentrations and RBC K^+ depletion showed pronounced delay between maximal digitoxin plasma concentrations and maximal RBC K^+ depletion (Urtizberea 1991). Plotting the plasma digitoxin concentrations as a function of effect gave a anticlockwise hysteresis loop for the six patients. Figure 1 presents the effect-concentration curve for one patient. A similar delay had been previously described for the assessment of correlations between haemodynamic changes characterizing inotropic effect

and plasma levels (Shapiro 1970; Kramer 1979). Ochs (1981) showed that after infusion of digitoxin, the ejection fraction increased 4 to 8 hours after the infusion. Kramer (1979) correlated the pharmacokinetics of digoxin in the peripheral tissue compartment and the time course of the electrochemical systole QS_2 in patients under digoxin therapy. In digitoxin-intoxicated patients, no relationships between RBC K^+ depletion and predicted tissue levels of digitoxin have been described because of the existence of a delay between Ct_{max} and E_{max}.
So, the effect concentration at the action site of digitoxin was calculated using the effect compartment of Sheiner and was correlated with the RBC K^+ depletion using the E_{max} sigmoid model (Fig. 2). Keo values ranged from 0.01 to 0.025 h^{-1} (0.015 ± 0.005 h^{-1}) and the EC_{50} values ranged from 36 to 112 ng/ml (70.9 ± 22.6 ng/ml). The validity of the E_{max} sigmoid model was confirmed by the good fit between the calculated effect versus time curve and the experimental data (Fig. 3). The mean EC_{50} that we reported was in the same range as the *in vitro* values found for digitoxin inhibition of RBC ^{86}Rb uptake by erythrocytes by Hamlyn (1989).
Bradycardia is observed 3 hours after intoxication by cardiac glycosides and may be investigated in TK-TD modelling. Recently, Vetticaden (1989) have observed a delay between the onset of digoxin-induced bradycardia and maximal plasma concentrations of digoxin in dogs. They used the effect compartment of Sheiner and the E_{max} sigmoid model. Scherrmann also found this model was adequate in modelling digitoxin kinetics and bradycardia in a patient intoxicated by digitoxin with an EC_{50} of 77 ng/ml (unpublished data) .

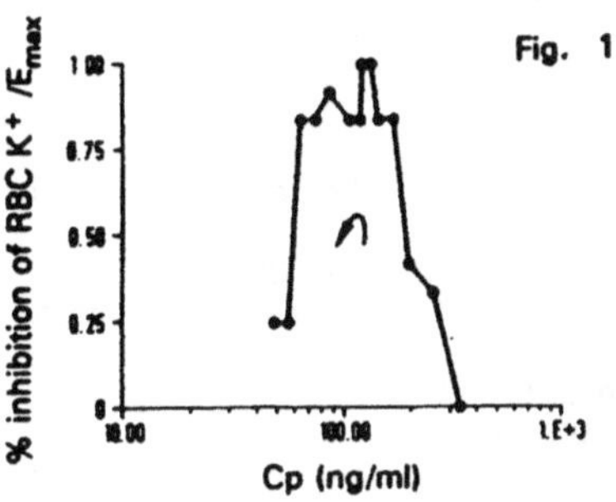

Fig. 1

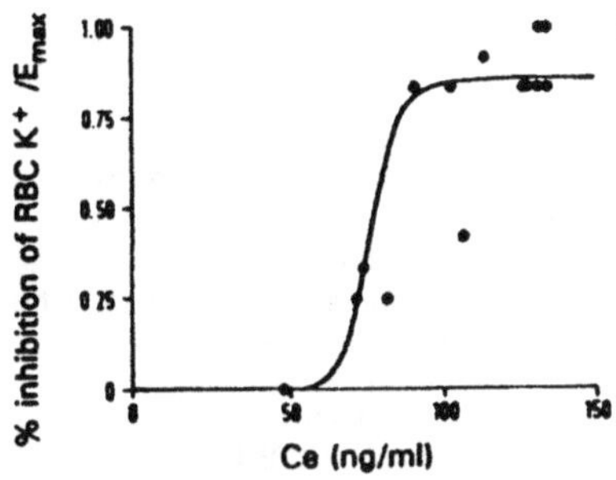

Fig. 2

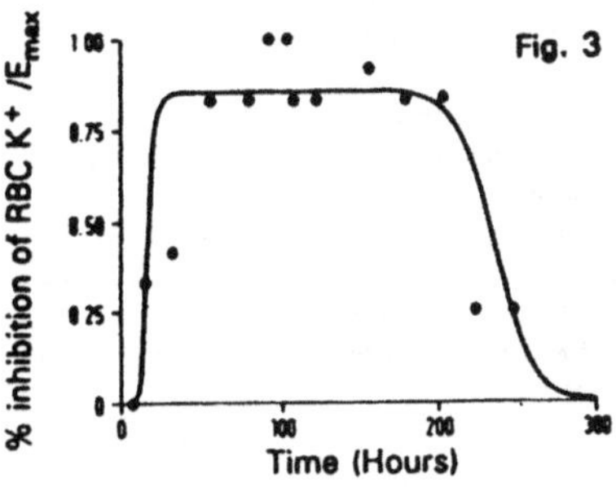

Fig. 3

Figure. 1 Relationship between RBC K^+ depletion and plasma concentration of digitoxin in one digitoxin-intoxicated patient (anticlockwise hysteresis)

Figure. 2 Relationship between RBC K^+ depletion and digitoxin effect compartment concentration (sigmoid E_{max} model)

Figure. 3 Kinetics of RBC K^+ depletion. The fitted curve was generated by using sigmoid E_{max} model.

LIMITS TO PK-PD MODELLING IN CLINICAL TOXICOLOGY

Actual limits for PK-PD or TK-TD modelling of drugs are firstly linked to the measurement of the effect. Very often, the argument that the effect cannot be measured is used to reject PK-PD studies. Which effect has to be measured? Whenever possible, direct numeric recording of PD measurement have to be preferred (e.g. blood pressure, heart rate, electrocardiographic recording). But, pharmacodynamic effects of drugs cannot be easely measured. For example, analgesic effect of drugs evaluated by questioning the patients is highly dependent on the patient's appreciation. Attribution of the effect to parent molecule has to be checked. If not, the pharmacokinetics of the active metabolites has to be determined and must be taken into account for PK-PD modelling. Determination of Eo is not always easy. Expression of the effect in relation to maximal effect has to be done to allow comparison between the different subjects of the same study. Moreover, assessment of relationships between toxicokinetic and toxicodynamic parameters is often difficult in clinical toxicology. In contrast to classical pharmacological practice, under emergency conditions blood sampling cannot always be performed at the same time after drug ingestion for all patients and several parameters such as ingested dose and time of ingestion cannot be accurately verified with self-poisoned patients. Moreover, the distribution phase cannot be observed if blood sampling is performed several hours after drug ingestion. In these cases, toxicokinetic-toxicodynamic modelling may be performed using only the elimination phase but is not possible for drugs whose pharmacodynamic effect occurs during absorption or distribution phases (Nieldsen-Kudsk 1988).

CONCLUSION

A model based approach to establishment of relationships between toxic effects and pharmacokinetics of toxins would be of interest in clinical toxicology. The classical methods of PK-PD modelling are applicable for toxic effects of drugs in acute self-poisonings as it has been performed for cardiac glycosides. From these data, the occurence of toxic effects could be predicted in function of first plasma concentrations of the toxin.

REFERENCES

Bismuth C, Gautier M, Conso P (1973) Hyperkaliemia in acute digitalis poisoning: prognostic, significance and therapeutic implications. Clin Toxicol 6: 153-162

Grevel J (1987) Kinetic-effect models and their applications. Pharm Res 4: 86-92

Hamlyn JM (1989) Increased levels of a humoral digitalis-like factors in deoxycorticosterone acetate induced hypertension in the pig. J Endocrinol 122: 165-166

Henion WA, Montodo A, Hilala A, Bayer R, Cohen J (1983) Changes in red blood cells electrolytes concentrations in digitalis intoxication. 106: 14-20

Holford NHG, Sheiner LB (1981) Understanding the dose effect relationship: clinical application of pharmacocinetic-pharmacodynamic models. Clin Pharmacokin 6: 14-20

Kelman AW, Whiting B (1980) Modeling of drug response in individual subjects. J Pharmacokin Biopharm 7: 47-61

Kramer WG, Kolisbash AJ, Lewis RP, Bathala MS, Visconti J, Reuning RH (1979) Pharmacokinetics of digoxin relationship between response intensity and predicted compartment drug levels in man. J Pharmacokin Biopharm 7: 47-61

Levy G (1966) Kinetics of pharmacologic effects. Clin Pharmacol Ther 7: 462-472

Loes MW, Singh S, Lock JE, Birkin M (1978) Relation between plasma and red-cell electrolyte concentration and digoxin levels in children N Eng J Med 299: 501-504

Nielsen-Kudsk JE, Askholt J (1988) Pharmacokinetics of enprofylline and theophylline in the isolated rabbit heart. Pharmacol Toxicol 63: 193-188

Ochs H, Grube E, Greenblat DJ, Arendt R, Bodem G (1981) Pharmacokinetics and pharmacodynamics of intravenous digoxin and digitoxin. Klin Wschr 59: 889-897

Shapiro W, Natahara K, Taubert K (1970) Relationship of plasma digitoxin and digoxin to cardiac response following intravenous digitalization in man. Circulation 49: 1065-1072

Sheiner LB, Stanski DR, Vozeh S, Ham J (1979) Simultaneous modeling of pharmacokinetics and pharmacodynamics: application to d-tubocurarine. Clin Pharmacol Ther 29: 358-371

Urtizberea M, Rochdi M, Baud FJ, Scherrmann JM (1990) Toxicokinetic-toxicodynamic models describing the relation of plasma and red blood cells potassium with plasma digitalis acute in acute human digitalis poisoning. Tox in Vitro 4: 526-531

Urtizberea M, Rochdi M, Sabouraud A, Bismuth C, Scherrmann JM (1991) Relationship between red blood cell potassium and plasma digitoxin concentrations in intoxicated patients Pharmacol Toxicol 68: 237-242

Vetticaden SJ, Barr WH, Allison TB (1989) Pharmacodynamic modeling of digoxin induced bradycardia. J Pharmacokin Biopharm 17: 89-107

Bioanalysis in Toxicology

(Chair: F. A. de WOLFF, Leiden)

Analytical Forensic Toxicology

M David Osselton

Central Research and Support Establishment, The Forensic Science Service, Aldermaston, Reading, Berkshire, RG7 4PN, UK.

My brief for this paper is to provide an overview of forensic toxicology and to illustrate what it is about forensic toxicology that makes it unique. I would therefore like to outline the background in which the forensic toxicologist works, describe some of the sorts of problems which may be encountered and illustrate how modern advances in scientific methodology may be applied to assisting the forensic toxicologist fulfil his objectives.

The use of poisons is as old as humanity and poisons in the form of various plants, minerals and synthetic materials have been used for fishing, hunting and the destruction of animal and human enemies. Poison has been the murderer's friend for thousands of years and is the most subtle and premediated way of committing the ultimate crime. Throughout the ages great knowledge has been accumulated on the actions and use of poisonous substances. Early Indian and Egyptian writings included references to poisons and their antidotes whilst Greek and Roman literature has reported the use of poisons and venoms for suicidal and homicidal purposes. As far back as 339 B.C. Socrates was executed by being made to drink an extract of hemlock. An Arab alchemist named Jabir became the first man to prepare arsenious oxixde, a white, tasteless and odourless powder that apparently left no traces in the body. Jabirs preparation seemed the ideal poison and became a favourite murder weapon in the middle ages because its symptoms - violent stomach aches and vomiting - resembled those of cholera (a common disease of the times). During the Renaissance, poisoning became an art and European court histories are filled with accounts of deaths by poisoning of kings, popes and nobility. The problem became so bad that guards were required to eat the food and drink the wine of their masters as a means of detecting the presence of poisons. The forensic toxicologist of today has, fortunately, advanced somewhat in his analytical techniques and is no longer required to implement the use of taste testing.

Forensic science may be defined as the application of science and scientific techniques to investigating crime, with a view to providing the law courts with impartial evidence bearing on the issues before them. Forensic toxicology is that area of forensic science which deals specifically with the application of scientific techniques for the detection and analysis of drugs and poisons and as such may involve

(a) The identification of drugs of dependence in criminally related drugs offences e.g. the possession of controlled substances or the investigation of clandestine laboratories.

(b) The isolation and identification of drugs or other poisons from body tissues in cases of suspected fatal poisoning.

(c) The analysis of alcohol and other drugs in body fluids taken from living subjects in driving, murder or rape cases.

In the latter sorts of cases the compounds most likely to be implicated are those which exhibit some effect on the central nervous system and those most frequently encountered include tranquillisers, sedatives, hypnotics, narcotics and hallucinogens. In the murder case a suspect or accused person frequently claims to have been under the effect of a drug when the crime was committed and may imply that he/she was therefore not responsible for his/her actions. Rape victims frequently claim that they have been drugged prior to an alleged assault but one of the biggest problems is associated with the involvement of drugs in driving accidents or deaths.

Recent surveys of drivers have shown that when tested for the presence of drugs which can affect driving ability, 26% were positive in Ontario, 25% in California (1969), 24% in New South Wales, 21% in California (1968), 8.4% in Melbourne and 5.5% in Belfast. Such surveys indicate that driving under the influence of drugs is a major international problem and one which is likely to stay with us for some time to come.

The birth of analytical forensic toxicology may perhaps be attributed to Plenck after his statement that "The only certain sign of poisoning is the chemical identificaton of the poison in the organs of the body." A number of advances in analytical toxicology were made during the late 18th century and 19th century which included the classification of poisons and the first extraction of arsenic from tissues by Orfila, methods for the detection of arsenic by Scheele and Marsh and the development of methods for the isolation of alkaloids from human organs by Stas and Otto.

Whilst the work of the early forensic toxicologist was largely confined to the analysis of hemlock, aconite, opium and toxic metals such as arsenic the number of potential poisons which are available today is vast. The present day forensic toxcologist may be required to analyse for many hundreds of different substances during the course of his work. The aim of the forensic toxicologist is to provide information relating to drug or poison concentrations which may be used to explain post mortem, clinical or behavioural effects. In addition to being a competent analyst, the forensic toxicologist must therefore also be able to predict the likely response of the individual to any drug/poison concentration and, where necessary, use pharmacokinetic information to help him to interpret his results and reach his conclusions.

Much of the work of the forensic toxicologist is now concerned with the analysis of organic substances in body tissues rather than the inorganic poisons used in earlier times. Inorganic poisons such as arsenic, antimony and cyanide are now tightly regulated and their supply is controlled making them difficult to obtain. Although garden and agricultural chemicals are encountered from time to time the incorporation of coloured or foul smelling reagents into these chemical preparations makes their undetected use by the criminal poisoner difficult. The restricted availability of poisons does not mean that they are unavailable however and an open mind must be kept at all times since many poisons such as cyanide are used industrially. I recall the case of an Asian gentleman who, whilst preparing himself a drink of tea, was almost poisoned by cyanide. It was his practice to boil a pan of water and then add the tea, milk and sugar to the boiling water and then to simmer the mixture for a few minutes on the stove. Whilst performing this procedure he experienced a tightening of his chest muscles, chest pain and a choking sensation. In view of an earlier dispute which had taken place with his brother in which his life had been threatened, he suspected that an attempt had been made to poison him and so the incident was reported to the police. The police visited the home of the complainant and were treated to a demonstration of the tea making procedure in the very small kitchen. After the milk and sugar had been added to the tea and boiling water, the policemen started to experience chest pains and difficulty in breathing and so samples of the milk, sugar and tea were submitted for laboratory analysis. On examination, cyanide could be clearly smelled in the milk and sugar and analysis revealed a total of 32 grams of sodium cyanide to be present. The complainants brother, who was a printer, was arrested and found to be in possession of nearly one kilogram of sodium cyanide which he had apparently obtained from his work place. Since many industrial processes utilise toxic materials, background information about a case such as the occupation of the victim or his/her relatives can provide vital clues which can serve to reduce time spent on analysis.

Ten to fifteen years ago the number of organic poisons encountered was relatively small and in most cases the doses required to achieve fatal poisoning were relatively large. During recent years we have witnessed a dramatic change of emphasis in forensic toxicology. The number of straight forward cases have almost disappeared and instead, we are required to turn our skills to solving deaths with complex drug mixtures, drug driving cases and the analysis of blood from suspects accused of rape or murder who claim that the felonies of which they were accused were performed under the influence of drugs taken before the offence was committed. Recent trends in pharmaceutical development have been towards producing potent, low dose drug preparations as may be exemplified by looking at some of the old and new hypnotic and anti-anxiety drugs. Whereas many of the older pharmaceutical compounds which were used because of their psychoactive properties required intake of several hundreds of milligrams per day to achieve a therapeutic effect, many modern compounds are taken in doses between 0.25 and 10 milligrams per day. This is reflected in the circulating blood levels and highlights that even with todays improved analytical technology, it is no easy task to make reliable measurements. Unlike our pharmaceutical or clinical colleagues, the forensic toxicologist has to work with "one off" cases where there is little or no control over the samples submitted for analysis.

The clinical toxicologist's major task is to monitor compliance with therapy and to relate measured plasma levels of drugs with drug treatment using pharmacokinetics. The forensic toxicologist uses pharmacokinetic data retrospectively in an attempt to explain whether drugs may have affected behaviour some time prior to the blood sample being taken when a criminal offence allegedly took place. The clinical toxicologist also frequently has several data points on which to base his calculations and usually has access to accurate and precise information concerning the drug taken, the dosage and time. He or she is also in a position to go back for more samples or even repeat the experiment. In forensic toxicology, we usually have one small sample on which all screening, qualitative and quantitative analyses have to be carried out. Furthermore the information provided about what poison may be involved is usually incomplete and on occasions is totally erroneous. There is no opportunity to go back for a second sample from a living subject and more often than not the forensic sample is rarely ideal. Blood specimens are invariably haemolysed and tissue samples may frequently be well into the stages of post mortem decomposition. In forensic toxicology, we rarely have the privilege of working with nice clean plasma samples; instead, we have to contend with extracting compounds of interest from a complex biological matrix. Not only does this complicate the clean-up procedures during the extraction phase, it also complicates the interpretation of the results since little has been published concerning the distribution of drugs in whole blood.

Samples submitted for forensic examination are regarded as the property of the court or a coroner and must therefore be accounted for at all times. Each sample must be clearly and precisely labelled and documented, and a complete record must be maintained of its movements. This is generally referred to as the "chain of custody" since it is vital that the court can be satisfied that the sample was not tampered with or altered in order to falsify evidence.

In addition to the unique case circumstances which have to be considered in forensic toxicology, the site and mode of sample collection may have significant implications to a case. Blood taken from a living subject is usually collected from the cubital vein and may be stored in containers containing fluoride as a bactericidal agent if alcohol is to be measured, or alternatively in heparinised containers for general drug analysis. The site of sampling for post mortem blood samples is more critical and should, where possible, be from the femoral region which is well away from the major body organs. Whilst the heart may provide the pathologist with a readily available pool of blood, it is unsatisfactory for forensic purposes since many drugs may diffuse into the heart from the liver via the major blood vessels after death. In such cases the resultant analysis could lead to very erroneous interpretation. The containers into which samples are collected can also exert a profound effect on drug analysis and the subsequent interpretation of results. For example, it has been shown that the plasticisers in "vacutainer" syringes may cause displacement of protein bound drugs leading to misinterpretation of results. Plasticisers found in containers or rubber septa can also cause gross interference with chromatographic procedures used for drug identification.

Because of the wide range of poisons which may be encountered in forensic toxicology, it is convenient to classify substances into seven groups viz. gases, volatile substances, drugs, metals, pesticides, anions and miscellaneous substances. Unfortunately, there is no simple or universal test for a poison and so the basic approach is to employ a battery of simple screening tests for each of the major groups of poisons after which more detailed analysis is performed.

Simple screening tests may involve spot tests or immunoassays since these use a minimum of sample and are usually quick and inexpensive to perform. The major drawback to such tests however is that they are not specific and have to be followed up by more precise procedures.

Once a drug or poison has been implicated by screenning it is necessary to extract it from the tissue and to purify it prior to analysis. Organic solvents are commonly used for extracting drugs from tissues after the tissue has been suitably broken down to yield a workable matrix. Whilst direct solvent extraction is suitable for isolating drugs from blood, solid tissues require some form of pretreatment to breakdown the tissue to release drugs into a matrix from which they can be extracted. Traditionally this has been achieved by either acid hydrolysis or the addition of a protein precipitating reagent however both of these methods suffer from disadvantages. Whilst acid hydroloysis may breakdown much of the protein structure of the tissue, many drug substances may also be degraded. Protein precipitation techniques can result in the co-precipitation of protein bound drugs with the proteins to which they are bound. In both cases, significant drug losses can occur. More recently, proteolytic enzymes have been used to break down tissues under very mild conditions resulting in the release of much higher levels of drugs for analysis. Proteolytic enzymes are particularly suitable for cases where acid labile or protein bound drugs are involved and have also proved to be useful for the analysis of inorganic poisons.

One particularly interesting example of the use of proteolytic enzymes has been for the analysis of drugs in soil following the exhumation of a corpse several years after burial. The young lady at the centre of the story was in the habit of living away from her parents home for long periods of time and on the occasion in question had turned up unexpectedly to see her widowed mother. During the course of her visit, she had been involved in a disagreement with her mother and shortly after the disagreement, her mother went out shopping. The young lady informed her young sister that she believed in reincarnation and thought that it was about time that she deserved a better life. She told her younger sister that she was going to go out into the garden, dig herself a grave and take an overdose of sleeping tablets. She then asked her younger sister to bury her by covering her up with soil after she had died. This, the younger sister agreed to do. When the mother returned home, she asked the younger sister where the elder sister was, to which the younger sister replied "she's gone" meaning that she was dead. The mother interpreted this as meaning that the elder sister had left home again and therefore thought nothing more about it. Some two days later whilst walking in the garden, the younger sister came across her elder sisters body in a shallow grave and remembering her promise duly covered the corpse with soil. Nothing more was said about the incident until seven years later when a notice was received that the house was to be demolished to make way for the construction of a new road. It was at this point that the younger sister informed her mother that if the garden was to be dug up, the workmen would come across a body. You can imagine the surprise of the local police sergeant when the two women walked into his police station and recounted this bizarre story. The police were duty bound to follow up the story and sure enough, the body of a young female lying in a prone sleeping position was uncovered in the garden. The task of the laboratory was to verify whether the story which had been told to the police was feasible and it was therefore our task to determine whether any traces of drugs could be identified. The body was completely decomposed and all that remained was a clothed skeleton so it was decided to collect a sample of the soil from under the abdominal region of the corpse. During excavation of the body, an empty tablet bottle was found close to the head and further examination of the inside of the bottle indicated that it had contained butobarbitone. Conventional methods of analysing the soil revealed no traces of drug substances however it was at the time when the use of subtilisin Carslberg had been discovered to provide increased levels of drugs from tissues so, I was requested to try out the new technology on the soil sample. The soil was treated first with a mixture of cellulase and hemicellulase enzymes to break down plant materials and then by subtilisin Carslberg digestion. To our surprise, we were able to detect both butobarbitone and its metabolite hydroxybutobarbitone in the soil sample.

The quest for improving analytical ability is a continuous process in all areas of toxicology and there are still times when low levels of drugs can have a significant implication. One of the current problems we face is concerned with the ability to detect, quantify and confirm the presence of low nanogram quantities of opioid drugs in small blood samples. During recent months we have been investigating the concept of antibody medicated extraction which involves passing a small sample of blood down a column of immobilised antibodies. The philosophy behind the technique is that the immobilised antibodies will specifically bind drugs against which the antibodies have been raised whilst all other substances present in the sample pass through the column. This theoretically allows a specific single step extraction and clean up procedure. Results to date have been very promising and yields of greater than 90% have been obtained from spiked drug samples. The extracts are exceptionally clean and produce chromatographic traces which are free from endogenous interferences and background noise thus permitting trace quantities of opioids to be detected.

Accompanying developments in extraction technology have been significant advances in instrumental technology. Capillary column gas chromatography provides high resolution of substances on-column and this, together with specific chromatographic detectors increases our ability to detect low levels of drugs in tissue extracts. One of the prime guide-lines followed in forensic toxicology is that substances must be analysed using at least two techniques which are based on different principals. This is to overcome the possibility of misidentification of a chromatographic peak. The chances of two totally different substances co-chromatographing on two non-correlating systems are less than if only one system is used and are reduced still further if three systems are employed. For this reason a combination of gas chromatography with group specific detectors, HPLC and mass spectrometry are commonly employed.

The interfacing of HPLC and mass spectrometry still has its problems for routine application to forensic toxicology but the development of the diode array detector has made a significant impact on our work. The diode array detector enables the HPLC eluent to be scanned at several different wavelengths simultaneously to detect compounds with different absorption spectra and this, linked with modern data handling offers us the facility of displaying and plotting the spectra of eluting compounds and of comparing them with reference library spectra.

The use and development of analytical techniques, like the development of new and more potent pharmaceuticals, is a continuous process and it is essential that we keep abreast and aware of new developments in both areas. I believe that supercritical fluid chromatography (SFC) and supercritical fluid extraction (SFE) are both areas which are likely to exert a profound impact on our work in the future. Although the principals of SFC have been known for many years, we are only now at a point where the apparatus is becoming commercially available. SFC offers many of the potential advantages of HPLC and GC in one technique. Chromatographic conditions are mild and both capillary and packed column technology can be used. The nature of supercritical fluids enables columns to be connected to a range of chromatographic detectors e.g. flame ionisation, phosphorus nitrogen detectors, ultraviolet and mass spectrometric. The combination of SFE with SFC may in the future provide us with a single stage extraction and analysis procedure "on-line". Another new and potentially exciting technique is High Performance Capillary Electrophoresis (HPCE), a procedure which offers a combination of high speed, high resolution and high sensitivity for the analysis of nanolitre volume samples. The technique involves the electrophoretic separation of samples within a capillary tube under the influence of high voltage and may be potentially applicable for the analysis of many substances of interest to the toxicologist. Although still in its infancy, HPCE could offer us ways of screening for the presence of drugs, protein and peptide toxins and even anions and cations. Since the technique is claimed to be capable of differentiating between proteins or peptides which differ by only one or two amino acids it could have important forensic implications. For example, in cases of insulin poisoning the technique has the potential to separate, and hence identify, human insulin, synthetic insulin and insulin from different animal species and could therefore help us to identify the source of insulin used.

With the host of methods now available to the analytical toxicologists, careful consideration is required in the selection of a method for a particular purpose. Methods should be good enough for the purpose to which they are put, without being over costly, however it is important to ensure that they will do the job for which they are required. Only after careful consideration of the known facts in a case can an appropriate method be selected.

This may be exemplified by the case of a child who consumed a bottle of enteric coated aspirin tablets in mistake for sweets (they looked like a common brand of children's sweets marketed as Smarties). The child's mother took the child to hospital together with a sample of one of the tablets which had been ingested. A stomach wash was performed but nothing was recovered in the way of tablet material. A blood sample was sent to the hospital laboratory for salicylate analysis and the child placed under observation. The laboratory tested the blood sample for salicylate using the Trinders Test and obtained a negative result. The child showed no indication of illness and so, after a few hours was sent home. Early the following morning, the child became very ill and died.

Analysis of the post mortem blood showed the presence of exceptionally high levels of salicylate which could account for fatal poisoning. We were later requested to analyse the post admission blood sample to ascertain whether the hospital laboratory had made a mistake and was therefore negligent. Using HPLC, there were clearly indications that salicylate was present in the blood however, the levels of salicylate were below the level of detection limit of the Trinders Test. Whilst this vindicated the hospital laboratory of negligence it does illustrate that it is necessary to know the limitations of the methods being used. Whilst the Trinder Test may be adequate for the screening of samples taken from a comatose patient admitted as an overdose victim in a casualty department it is of less use for screening where low or even subtherapeutic quantities of salicylate may be present.

Whatever methods are used in forensic toxicology, it is vitally important that they are robust, legally acceptable and can be defended in a court of law. It is also important that we know what our methodology is capable of achieving. Of the numerous methods published for the analysis of drugs in human tissues few have been fully characterised so that the user can be confident of the accuracy, precision, reproducibility and robustness, sensitivity, linearity and limitations of the assay.

Although the nature of forensic toxicology work is unique, it is an area of toxicology which is profoundly influenced by what is happening elsewhere in the scientific world. Most importantly, forensic toxicologists have now agreed through the International Association of Forensic Toxicologists the need for a course of action towards getting their methodology and working practices more formalised.

In 1992 many new frontiers will be opening in Europe and there will be opportunities for laboratories to expand their work across borders. Those who wish to do so however will have to be able to prove their credibility and operate to criteria which can be internationally recognised. The Forensic Science Service places great emphasis on quality and is now actively working towards the formalisation of GLP, the full documentation of Standard Operating Procedures and Accreditation. The critical examination of methods and procedures will be one of the big challenges facing all toxicologists in the immediate future. The purpose of analytical toxicology is to produce information which can be interpreted and although we may have little control over the many and varied types of materials and samples we are required to analyse, we do and must have absolute control over the methods we use. Only when we become fully aware of our analytical capabilities will we be able to advance the science of analytical toxicology to its full potential.

Biological Effect Monitoring

C.T.A. Evelo and P.Th. Henderson

Department of Occupational & Environmental Health and Toxicology,
University of Limburg, PO Box 616, 6200 MD Maastricht, The Netherlands.

Introduction

Monitoring activities have an important role in occupational and environmental practice. Techniques to detect and to control chemical exposure in order to protect people from environmental and occupational illness are rapidly expanding. Most obvious are the methods related to the evaluation of the presence of xenobiotic chemical agents either in the contaminated environment or in the exposed organism. In the first case the procedures are indicated as *environmental monitoring* (EM). In inhalatory exposure, estimations can be made of the 'intake' or 'external load' on the basis of known concentrations in the ambient air. The actual 'uptake' or 'internal load', however, can only be approximated by means of EM methods. On the other hand, measurement of the chemical agents and/or their metabolites in tissues, secreta or excreta of the exposed organism provides an estimate of the total uptake. The latter approach has been defined *biological monitoring* (BM) (Berlin 1984). In addition to these methods many other preventive activities have been

Table 1: Overview of monitoring methods in occupational and environmental toxicology.

	Monitoring type:	Assessment of:
EM	Environmental Monitoring	**external** exposure
BM	Biological Monitoring	**internal** load
BEM	Biological Effect Monitoring	**biological effects** (not necessarily health related)
HS	Health Surveillance	**health** state

developed, including periodic medico-physiological examinations of the exposed people. This apparoach is more directly related to the detection of early health effects, and is defined as *health surveillance monitoring* (HS). Changes in physiological functions of the organism can be indicative of changes in the state of health as a consequence of a certain chemical exposure. Such information on the state of health or early impairment of health is very useful in the prevention of overt intoxications due to continuation of overexposures. Well-known examples in this area are the increased levels of serum transaminases providing information on hepatotoxic effects, e.g. after exposure to certain halogenated alkanes. Most of the HS parameters a priori have to be regarded as adverse effects, which means that their significance in terms of prevention is limited. 'Early' adverse effect should also be prevented. Generally, an impaired state of health of the organism can be considered as

Table 2: Causality chain leading to health effects.

Chemical contamination of the environment
⇓
External exposure (intake)
⇓
Internal exposure (uptake, distribution, metabolism, excretion)
⇓
Primary effects at molecular, biochemical and cellular levels.
⇓
Changes in physiological functions
⇓
Adverse effects, health impairment, occupational/environmental illness.

an endpoint of a sequence of several events at lower levels of homeostasis. Such a 'causality chain' can be depicted as shown in table 2.

In recent years the arsenal of monitoring techniques has been extended with test-methods directed to the assessment of early, possibly reversible, biological effects (*Biological effect monitoring*: BEM). Mostly it is unknown a priori whether primary biological effects have to be regarded as adverse or non-adverse. In this respect the health significance of biological effect parameters has still to be established. In order to avoid confusion among the various monitoring programmes, Zielhuis and Henderson (1986) proposed the following definition:

> **Biological effect monitoring** (BEM) is "the measurement and assessment of early biological effects, of which the relationship to health impairment has not yet been established, in exposed workers to evaluate exposure and/or health risk compared to an appropriate reference".

Most of the tests for EM and BM are agent-specific or specified for certain groups of related chemical, e.g. thioethers, mutagens. The ultimate objective of EM and BM is to estimate the target dose. Because this dose or concentration in the critical organ is determined by toxicokinetic processes, (distribution, biotransformation, elimination), BM is based on 'the impact of organism on agent'. On the other hand, effect parameters (BEM, HS) refer to toxicodynamic processes and reflect the impact of the target dose on the state of health. Biological effects may be due to many causes: chemical, non-chemical as well as multifactorial causes may be involved. Therefore, BEM and HS are generally agent-nonspecific.

Example: Effects on the Glutathione Dependent Detoxification System in Erythrocytes

Some aspects of the BEM approach can be examplified by methods that are based on changes in the glutathione system in erythrocytes after exposure to electrophilic agents.

Glutathione Depletion

Depletion of liver glutathione (GSH) during exposure to chemically reactive compounds is a well known phenomenon both *in vitro* and *in vivo*. GSH is also involved in the metabolism of many substances that pose an occupational health hazard. This leads to the urinary excretion of thioethers (Henderson *et al* 1984), which itself is used for biological monitoring. In principle, exposure to such compounds might lead to a decrease in GSH concentrations not only in the liver but also in more easily accessible material like blood cells. *In vitro* GSH in erythocytes can be depleted with thiol reactive substances like iodoacetamide, N-ethylmaleimide and diethylmaleate (Evelo and Henderson 1988) and also by metabolites of 3-acetamidophenol and cyclophosphamide (Palmen 1991). The reactivity of GSH is so high that it results in an effective protection of cellular macromolecules. Figure 1 shows that during *in vitro* treatment of erythrocytes with iodoaceetamide and N-ethylmaleimide the hemoglobin cysteine is not significantly alkylated as long as the intracellular GSH is not fully depleted. This kind of observations prompted us to study GSH levels in some situations where people are potentially exposed to electrophilic agents.

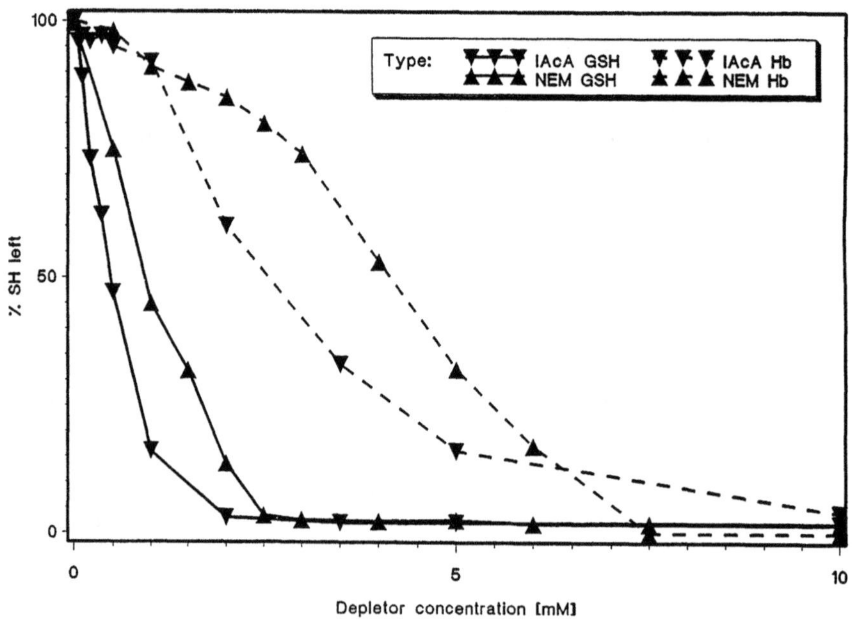

Figure 1: Free sulfhydryl groups of hemoglobin (Hb) and glutathione (GSH) after incubation of human blood with the depletors iodoacetamide (IAcA, 1 hour) and N-ethylmaleimide (NEM, 15 min.) at 37°C. Data are the mean of duplicate determinations. 100% HbSH values were 2.15 and 2.18 moles/mol and 100% GSH values were 0.86 and 0.72 mM respectively. Determined by titration with 4,4'-dithiopyridine (HbSH) and 5,5'-dithiobis 2-nitrobenzoic acid (GSH)

Glutathione S-Transferase Inactivation

The activity of erythrocyte glutathione S-transferase (GST ρ) is another interesting, and biochemically related, parameter. The activity of GST in general is known to be susceptible to many chemicals (Mannervik and Danielson 1988). The *in vitro* susceptibility of GST activity from human erythrocytes to industrially important electrophiles like acrolein, propylene-oxide, styrene oxide, ethylene dibromide and ethylene dichloride has been described by Ansari *et al* (1987). Kilpikari and Savolainen (1984) reported decreased values of erythrocyte GST activity in workers exposed to hot rubber fumes. Normal GST activity is known to vary largely between individuals (Strange *et al* 1982). In this respect the comparison of individual pre- and post exposure values is expected to be more sensitive than the comparison with a control group.

Creosote Workers

In a pilot study we determined GST activities in erythrocytes and GSH in blood of two workers creosoting wood. The samples for this study were collected as part of a biomonitoring study carried out by Jongeneelen and co-workers. Blood and urine samples of two exposed workers and four control persons, all non-smokers, were collected: at the beginning of a work week—which was after an exposure free period of several weeks—at the end of that week and after the weekend. Total (i.e. reduced plus oxidized) glutathione was determined with the cyclic oxidation reduction method (Anderson 1985). GST activity towards 1-chloro-2,4-dinitrobenzene (CDNB) was determined essentially as described by Habig and Jacoby (1981). Both methods were slightly modified as decribed previously (Brouwer *et al* 1991). Figure 2 shows the separate GSH and GST values for the two workers. In order to exclude errors caused by day to day analytical variations, values were expressed as the ratio of the individual value and the mean value of the four control persons determined the same day. In both workers GST and GSH values were decreased at the end of the week, and returned to values close to the initial values after the weekend. These data, which in themselves have a rather limited value, become more interesting when related to some other parameters determined in the same workers, as decribed below.

Environmental monitoring of pyrene and biological monitoring of urinary 1-hydroxypyrene, as markers for the exposure to polycyclic aromatic hydrocarbons, were done by Jongeneelen *et al* (personal communication). Environmental monitoring revealed that dermal and inhaled exposition for person 1 were generally at least two times higher than for person 2. Hydroxypyrene in urine was determined twice each day with previously decribed methods (Jongeneelen *et al* 1987). The individual excretion values for hydroxypyrene on the morning of the sampling days for the blood samples are shown in table 3. For both workers the urinary hydroxypyrene values were clearly above the normal range (normal 95 percentile value is 0.66 μmol/mol creatinine) at the end of the week. The interindividual differences in excretion were in accordance with the BM data.

Even more interesting were the findings by Steenwinkel and Baan (personal communication) who determined DNA-adducts of polycyclic aromatic hydrocarbons in white blood cells with the ^{32}P-postlabeling method developed by Randerath and co-workers (Randerath *et al* 1981; Watson 1987) in the same blood samples used for the GSH and GST

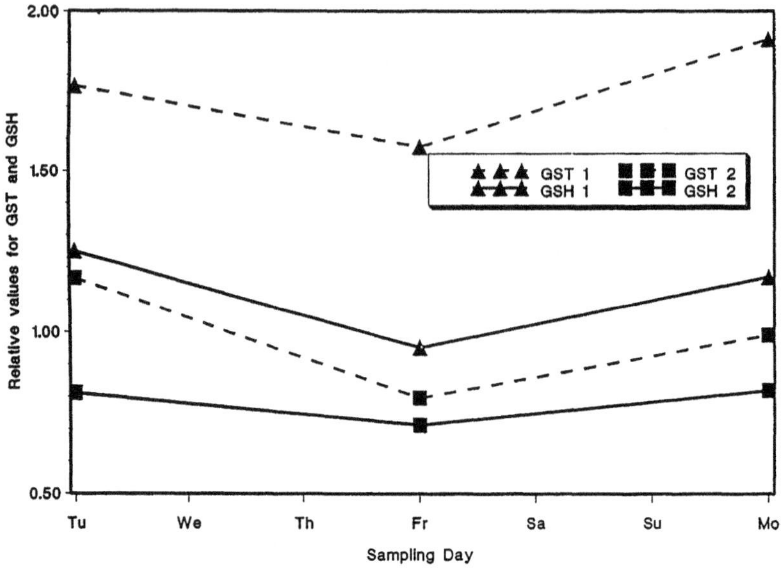

Figure 2: Glutathione concentrations in blood and Glutathione S-Transferase activities in erythrocytes of workers creosoting wood. Samples were collected before and after the work period and after the weekend. All values are the result of duplo determinations, and were expressed as relative values compared to samples from four non-exposed persons collected and determined on the same day.

determinations. The results are listed in table 3. All values are the means of two, in some

Sampling day	DNA-adducts [attomoles/μg DNA]		Hydroxypyrene exc. [μmol/mol creat.]	
	person 1	person 2	person 1	person 2
tuesday	6	n.d.[1]	0.35	0.70
friday	15	19	2.8	1.15
monday	13	18	0.55	0.5

[1] n.d. is not detectable, detection limit was 4 attomoles/μg DNA.

Table 3: DNA adduct values in creosote workers.

cases three, determinations. The analytical variation between experiments was about 20%. The end of week values are clearly elevated compared to the pre-exposure values. For the DNA-adducts a tendency for return to normal values is visible too, but this effect is small compared to the analytical variability. A slowness in recovery for DNA alkylation need not only be the result of slow repair processes, but can also be caused by continued availability of reactive compounds from internal storage in for instance fat tissue. Faster recovery of GSH and GST would, on the other hand, not only indicate the relatively fast reversibility of their impairment, but would also indicate that this impairment is caused by substances whose residency is relatively short.

Dichloropropene Workers

Further indications for the usefulness of the determination of GSH and GST came from a prospective study which was performed in the Dutch flower bulb culture, in collaboration with Brouwer and de Wolff, to investigate the possible effects of the soil fumigant 1,3-dichloropropene (DCP)(Brouwer *et al* 1991). At the start of the season in July, and after the season in October, blood and urine samples of workers applying DCP were collected. The activity of erythrocyte GST was significantly ($p<0.002$; Wilcoxon matched pair analysis) decreased from 4.7 before to 3.3 U/g Hb after the season. The GSH values in blood were also significantly ($p<0.02$) decreased from 0.93 to 0.82 mM. With a single exception for an unusual low GSH value in the pre-season samples, all individual values were found to be decreased at the end of the season (Fig. 3).

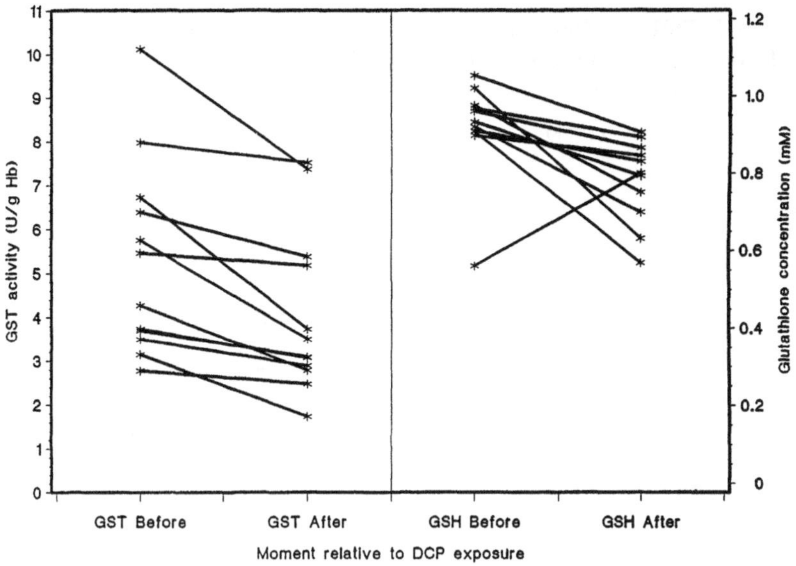

Figure 3: Glutathione S-transferase activity in erythrocytes and glutathione in blood of workers exposed to 1,3-dichloropropene.

The study was combined with environmental and biological monitoring of 1,3-dichloropropene (van Welie *et al* 1991). In the biological monitoring study excretion of two mercapturic acid metabolites of DCP was determined. Since these mercapturic acids are the result of GSH conjugation, their excretion indicates that GSH is actually used in the metabolism of DCP. Changes in health effect parameters for liver and kidney function in these workers were determined by Verplanke. Decrease of total serum bilirubin together with an increase in serum γ-glutamyltranspeptidase activity indicated moderate hepatic enzyme induction. And a possible subclinical nephrotoxic effect was indicated by significant increases in urinary excretion of albumin and retinol binding protein. The direction of the change in the determined health effect parameters for the liver—enzyme induction, not liver impairment—, and the size of the changes for the kidney functions—within the normal clinical range—, lead to the conclusion that there were no significant health effects on the individual level. Nevertheless, significant biological effects were found, which lead to the identification of internal exposure, and possibly increased risk.

Smokers

Since smokers are continuously exposed to electrophilic compounds originating from cigarette smoke, some effects on the glutathione system could be anticipated. Table 4 shows the concentration of total GSH in blood, erythrocyte GST activity and the amount of GST present in the erythrocytes. The latter amount was determined by a competitive ELISA assay with antibodies towards human GSTπ class enzymes (Medlabs, Dublin Ireland).

Table 4: GST and GSH in blood and urinary thioether excretion of smokers and non-smokers. Median values and ranges are shown. Symbols indicate significant differences from the values for non-smokers by Wilcoxon 2-sample test; *: $p<0.10$; ‡: $p<0.02$.

	non-smokers	smokers
age (mean ± s.d.)	33 ± 7 n=11	37 ± 10 n=10
GSH [mM]	1.06 0.88–1.32	1.12 0.77–1.17
GSTρ act. (CDNB) [U/ml blood]	0.71 0.42–1.13	0.52 0.31–0.95
GSTπ conc. (ELISA) ≡[U/ml blood]	0.6 0.3–4.5	4.2‡ 1.6–5.0
Thioether excr. [mmol/mol creat.]	3.3 0.5–9.9	4.7* 3.0–14.0

The thioether excretion in smokers was significantly increased. Such an increase is usually found in smokers (Henderson *et al* 1984), and it is included in the table only because it indicates exposure. Total blood GSH did not differ between smokers and non-smokers. The decrease in erythrocyte GST activity towards CDNB was not statistically significant. The amount of GST π class protein however, was strongly increased in the smokers. Thus a large fraction of the GSTπ class protein present in the erythrocytes of smokers is not active towards the common substrate CDNB. This may indicate that GST synthesis during erythrocyte formation is induced in smokers, to compensate for inactivation that takes place during the erythrocyte life-cycle. The induction of a GST not active towards CDNB would be another explanation.

Physical Activity

Since part of the GST activity loss may be caused by reactive oxygen species, we decided to study the influence high oxygen consumption due to physical activity (Evelo *et al* submitted). Previously sedentary men (n=23) and women (n=18) were trained to run a 1/2 marathon contest after 40 weeks. After 20 weeks of training total blood GSH and erythrocyte glutathione reductase (GR) activity were increased. After 40 weeks total blood GSH values were returned to normal, while GR remained elevated. Erythrocyte GST activity was strongly increased after 40 weeks, this may be indicative of the occurrence of lipid peroxidation in this training phase.

The participants ran a 15 km race after the first 20 weeks of training and a 1/2 marathon after 40 weeks. After the 15 km race total blood GSH and erythrocyte GST activity were decreased, while GR activity was increased. GSH and GR values were returned to before race conditions 5 days after the 15 km race. After the 1/2 marathon total blood GSH and erythrocyte GR activity were increased, while GST activity was decreased. 5 Days after the 1/2 marathon all values were returned to before race conditions. The difference in GSH sensibility after both races may be caused by differences in pre-race conditions. GSH was elevated before the 15 km race. The increases in GR activities after the races could at least partially be explained by concurrent increases in riboflavin concentrations.

Concluding Remarks

More and more methods in occupational and environmental toxicology, allow the determination of biological effects of which either no clear relationship to health impairment has been established, or at levels where no known clinical significance exists. The primary value of the use of such methods lays in the relation between the determined effects and the internal dose. If the relation between internal dose and the ultimate risk is known, like it is suggested for hemoglobin and DNA adducts (Henderson *et al* 1989), the assessment of biological effects can be used for risk estimations. However, even if such a clear relationship is not established the determination of biological effects may be indicative of exposure to reactive compounds, and may be used for the control of exposure situations. Much opposition against the use and development of BEM methods is based on a bad understanding of their results. Problems arise when health significance is attributed to results which do not allow such conclusions. It has often been overseen that methods leading to such results can still be very valuable for the management of risk. A clear distinction between BEM and HS is needed to improve this situation.

The development of new BEM methods depends largely on knowledge of toxicodynamics. The development of the methods for the glutathione system for instance, was stimulated by the understanding of the biochemical processes leading to thioether excretion and by insight in reactive oxygen toxicity. To put it in another way, knowledge of toxicodynamics can, and in our view should, be used for the development of new methods for biological effect monitoring.

Acknowledgements. The authors wish to thank Dr. Jongeneelen from the university of Nijmegen and Dr. Baan from the Medical Biological Laboratory of TNO for the opportunity to reference unpublished data on hydroxy-pyrene excretion and DNA-adducts in the creosote workers.

References

Anderson M.E. (1985) Determination of glutathione and glutathione disulfide in biological samples. *Meth Enzymol* **113**: 548–555.

Ansari G.A.S., S.V. Singh, J.C. Gan, Y.C. Awasthi (1987) Human erythrocyte glutathione S-transferase: a possible marker of chemical exposure. *Toxicol Lett* **37**: 57–62.

Berlin A., V. Riihimäki, H. Vainio (eds) (1984) Assessment of toxic agents at the work place. Roles of ambient and biological monitoring. Nijhof, Boston The Hague Dordrecht Lancaster.

Brouwer E.J., C.T.A. Evelo, A.J.W. Verplanke, R.T.H. van Welie and F.A. de Wolff (1991) Biological effect monitoring of occupational exposure to 1,3-dichloropropene: effects on liver and renal function and on glutathione conjugation. *Br J Ind Med* **48**: 167–172.

Evelo C.T.A. and P.Th. Henderson (1988) Influence of glutathione on the formation of cysteine alkylation products in human hemoglobin. *Toxicology* **52**: 177–186.

Evelo C.T.A., N.G.M. Palmen, Y. Arthur and G.M.E. Janssen (submitted for publication) Changes in Blood Glutathione Concentrations, and in Erythrocyte Glutathione Reductase and Glutathione S-Transferase Activity After Running Training and After Participation to Contests.

Henderson P.Th., R. van Doorn, Ch.–M Leijdekkers and R.P. Bos (1984) Excretion of thioethers in urine after exposure to electrophilic chemicals. In: Berlin A. *et al* (eds.) Monitoring human exposure to carcinogenic and mutagenic agents. IARC scientific publ. **59**, Lyon, p. 173–188.

Henderson R.F., W.E. Bechtold, J.A. Bond and J.D. Sun (1989) The use of biological markers in toxicology. *Crit Rev Toxicology* **20**: 65–82.

Jongeneelen F.J., R.B.M. Anzion and P.Th. Henderson (1987) Determination of hydroxylated metabolites of polycyclic aromatic hydrocarbons in urine. *J Chromatogr* **413**: 227–232.

Habig W.H. and W.B. Jakoby (1981) Assays for the differentiation of glutathione S-transferases. *Meth Enzymol* **77**: 398–405.

Kilpikari I. and H. Savolainen (1984) Decreased erythrocyte glutathione S-transferase activity in rubber workers. *Int Arch Occup Environ Health* **53**: 299–302.

Mannervik B. and U.H. Danielson (1988) Glutathione Transferases—structure and catalytic activity. *Crit Rev Biochem*, **23**: 283–337.

Palmen N.G.M., C.T.A. Evelo and P.Th. Henderson (1991) Depletion of erythrocyte glutathione by metabolites formed by fortified liver microsomes. Poster 154, presented at this Eurotox congress.

Randerath K., Reddy M.V., Gupta R.C.(1981) ^{32}P-Labeling test for DNA damage. *Proc Natl Acad Sci USA* **78**: 6126–6129.

Watson W.P. (1987) Post-radiolabeling for detecting DNA damage. *Mutagenesis* **2**: 319–331.

Welie R.T.H. van, P. van Duyn, D.H. Brouwer, J.J. van Hemmen, E.J. Brouwer and N.P.E. Vermeulen (1991) Inhalation exposure to 1,3-dichloropropene in the Dutch flower bulb culture. Part II. Biological monitoring by measurement of urinary excretion of two mercapturic acid metabolites. *Arch Environ Contam Toxicol*, **20**: 6–12.

Zielhuis R.L. and P.Th. Henderson (1986) Definitions of monitoring activities and their relevance for the practice of occupational health. *Int Arch Occup Environ Health* **57**: 249–257.

BIOLOGICAL MONITORING OF EXPOSURE TO BITUMEN FUMES DURING ROAD PAVING OPERATIONS

Sema Burgaz[1], Paul J.A.Borm[2], Frans J. Jongeneelen[3]

1Department of Toxicology, Faculty of Pharmacy, Gazi University, 06330, Ankara, Turkey.
2Department of Occupational and Environmental Health and Toxicology, University of Limburg, 6200 MD Maastricht, The Netherlands.
3Department of Toxicology, University of Nijmegen, 6500 HB, Nijmegen, The Netherlands.

INTRODUCTION

Bitumens and fumes contain relatively small amounts of polycyclic aromatic hydrocarbons (PAH), some of which are known to be carcinogenic or cocarcinonic (Brandt 1985). Workers may be exposed to bitumen fumes during transport, storage, production, handling, or use of bitumens (IARC 1985). There is great concern about health problems associated with occupational exposure to bitumen fumes, and among the suggested hazards is an increased frequency of malignancies (IARC 1985; Hansen 1991). Only a few researchers have performed biological monitoring of exposure to bitumen fumes during road paving operations (Triebig 1986; Lafuente 1987; Burgaz 1988; Pasquini 1989), however such studies are not conclusive as to the potential exposure to bitumens.

Current annual use of bitumens is about 0.35 million tons in Turkey. There is no available information on the extent of their exposure and health risks. The aim of the present study is to assess the exposure of workers to PAH during road paving operations with bitumens. To evaluate the PAH intake of the workers, the following urinary parameters are analysed: (i) urinary thioethers which is considered as an index of exposure to electrophilic compounds (Van Doorn 1981), (ii) urinary 1-hydroxypyrene as a marker of the internal dose of PAH (Jongeneelen 1988).

MATERIAL AND METHODS

Exposed workers were 39 males employed in road bituminization (mean age 35.5 years). Twentyone of the workers were smokers and thirtheen were receiving medication mostly analgesics and vitamins. Five of the workers had recently drunk alcohol. The majority of the exposed workers had worked on paving operations for at least two weeks prior to study. An occupationally nonexposed group of 29 men university staff and students (mean age 36.3 years) served as a control group. Fourtheen of the control subjects were smokers and nine were receiving medication mostly analgesics. One of the controls had recently drunk alcohol. All subjects had a normal diet. Most of the bitumens used in these applications were steam-refined and had 75 to 100 penetration values.

Urine samples from exposed subjects were collected at the end of working day on Thursday or Friday. Urine samples from control subjects were collected at different times of the day. After collection, the urine samples were divided into two aliquots to be used for the thioether assay (Seutter-Berlage 1979), and the analysis of 1-hydroxypyrene content (Jongeneelen 1987). Urinary thioether and 1-hydroxypyrene concentrations were corrected for creatinine. The creatinine content of the urine samples was measured with the method based on the reaction of creatinine with alkaline picrate (Donald 1967).

Since observed distributions of thioether and 1-hydroxypyrene excretion deviated from normality the values were transformed into logarithms. To evaluate the several types of possible influences on the biological parameters expressing PAH exposure five way analysis of variance was used.

RESULTS AND DISCUSSION

The results were given in Fig. 1 and Fig. 2. The mean of all the thioether excretions (SE) was 7.76 (0.75) mmol SH-/mol creatinine for exposed group and 4.61 (0.48) mmol SH-/mol creatinine for control groups (Fig. 1). Mean 1-hydroxypyrene excretions (SE) in the exposed group was 0.61 (0.06) umol/mol creatinine while mean 1-hydroxypyrene excretions in the control group was 0.28 (0.03) umol/mol creatinine (Fig. 2).

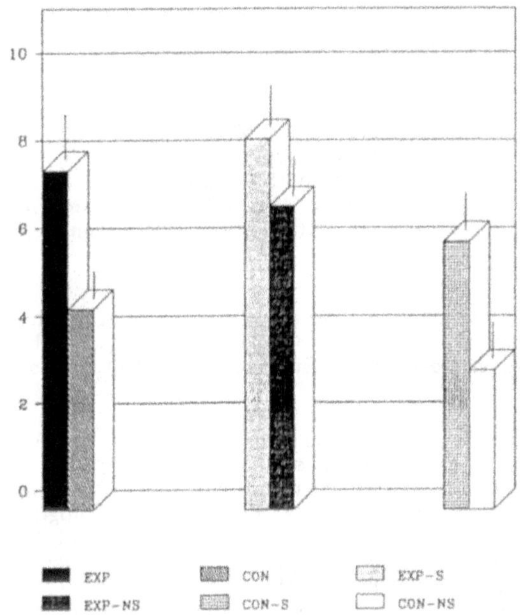

Figure 1. Mean values (SE) of urinary thioether excretion of exposed (EXP) and control (CON) groups, according to smoking habits (S=smoker, NS=nonsmoker).

Variance analysis indicates that exposure and smoking significantly ($p < 0.01$) contribute to the sum of squares for the thioethers and 1-hydroxypyrene. From Fig. 1 and 2, it is evident that bitumen fumes during road paving operations are absorbed by workers.

Increases in the thioether excretion may be related to exposure ($F=6.93$, $p=0.011$), but, the effect of smoking ($F=9.14$, $p=0.004$) was found to be stronger than that of exposure, medication seemed to influence the urinary thioethers ($F=3.76$, $p=0.057$). Compared to the effect of smoking, however, the effect of medication is slight. The effect of occupational exposure on 1-hydroxypyrene levels ($F=25.36$, $p=0.0001$) was stronger than that of smoking ($F=6.36$, $p=0.011$). Smoking increases the excretion of 1-hydroxypyrene, but it is not a strong determinant. Medication did not influence the 1-hydroxypyrene levels ($F=0.10$, $p=0.748$). Age and alcohol intake have no significant effect on both biological parameters.

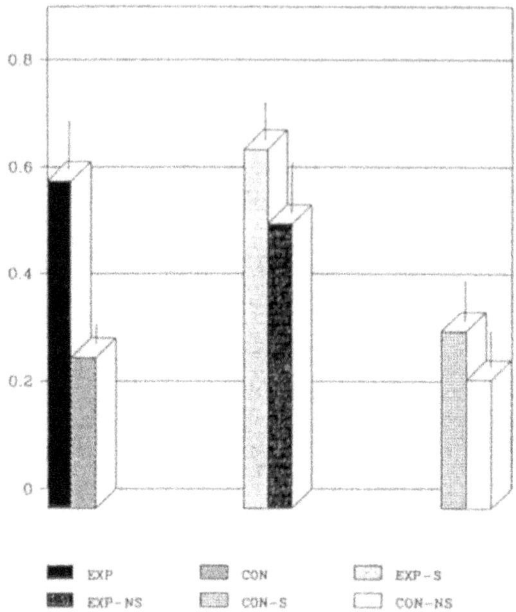

Figure 2. Mean values (SE) of 1-hydroxypyrene excretion of exposed (EXP) and control (CON) groups, according to smoking habits (S=smoker, NS=nonsmoker).

Despite of confounding variables such as smoking and medication with thioether assay, in agreement with the data previously published (Van Doorn 1981; Triebig 1986; Pasquini 1989), the present data suggest an exposure to electrophilic agents in bitumen fume.

Significant levels of 1-hydroxypyrene in road paving workers' urine reflect exposure to pyrene and ensure for a more detailed evaluation of exposure to genotoxic PAH in bitumen fume. In the present study and also in previous studies (Jongeneelen 1988, 1989) 1-hydroxypyrene has shown to be sensitive biological indicator of exposure to PAH. Moreover, our results show the applicability of the 1-hydroxypyrene assay after occupational exposure to petroleum-based products other than coal-tar derived ones. Since it is known that potential hazard is present to an appreciable extent in all paving operations with bitumens (Brandt 1985) it is essential to minimize exposure by controlling work method or personal protection.

REFERENCES

Brandt HCA, De Groot PC, Molyneux MKB, Tindlet PE (1985) Sampling and analysis of bitumen fumes. Ann Occup Hyg 29:27-80

Burgaz S, Bayhan A, Karakaya AE (1988) Thioether excretion of workers exposed to bitumen fumes. Int Arch Occup Environ Health 60:347-349

Donald M, Zimmer A (1967) Atlas of clinical laboratory procedures, vol 1. McGraw-Hill, New York

Hansen ES (1991) Mortality of mastic asphalt workers. Scand J Work Environ Health 17:20-24

International Agency for Research on Cancer (1985) IARC monographs on the evaluation of the carcinogenic risk of chemicals to humans. Polynuclear compounds, bitumens, coal-tars and derived products, shaleoils and soots, vol 35. IARC, Lyon

Jongeneelen FJ. Anzion RBM, Henderson PTh (1987) Determination of hydroxylated metabolites of polycyclic aromatic hydrocarbons in urine. J Chromatog 413: 227-232

Jongeneelen FJ, Anzion RBM, Scheepers PTJ, Bos RP, Henderson PTh, Nijenhuis EH, Veenstra SJ, Brouns RME, Winkes A (1988) 1-Hydroxypyrene in urine as a biological indicator of exposure to polycyclic aromatic hydrocarbons in several work environments. Ann Occup Hyg 32:35-43

Jongeneelen FJ, Anzion RBM, Theuws JLG, Bos RP (1989) Urinary 1-hydroxypyrene levels in workers handling petroleum coke. J Toxicol Environ Health 26:133-136

Lafuente A, Mallol J (1987) Urinary thioethers in workers exposed to the asphalt: An impairment of glutathione-S-transferase activity. J Toxicol Environ Health 21:533-534

Pasquini R, Monarca S, Scassellati Sforzolini G, Savino A, Bauleo FA, Angeli G (1989) Urinary excretion of mutagens, thioethers and D-glucaric acid in workers exposed to bitumen fumes. Int Arch Occup Environ Health 61:335-340

Seutter-Berlage F, Selten GCM, Oostendorp SGML, Hoog Antink JMT (1979) The modified thioether test. In: Strik JJTWA, Koeman JH (eds) Chemical porphyria in man. Elsevier, Amsterdam

Triebig G, Moser S, Schaller KH (1986) Pilot study for the diagnostic validity of urinary thioether excretion in occupational medicine. Zbl Arbeitsmed 36:301-307

Van Doorn R, Leijdekkers ChM, Bos RP, Brouns RME, Henderson PTh (1981) Detection of human exposure to electrophilic compounds by assay of thioether detoxication products in urine. Ann Occup Hyg 24:77-92

TOXICOLOGICAL INVESTIGATIONS ON UNUSUAL MATERIALS (HAIR AND VITREOUS HUMOR) : INTEREST AND LIMITATIONS.

KINTZ P., TRACQUI A., MANGIN P.

Institut de Médecine Légale, 11 rue Humann, 67000 Strasbourg, France.

Drugs and their metabolites can be easily detected in urine and plasma samples by classic analytical techniques (immunoassay or gas-liquid chromatography). Sometimes, these biological fluids are not collectable or are severly contaminated. Since several years, vitreous humor is used in forensic medicine for determination of time of death by using potassium estimations (Coe, 1969) or metabolic diseases (Dergun, 1986). Recently, some authors (Vogel 1981 ; Ziminski 1984 ; Caplan 1990) have proposed to make a toxicological screening in post mortem vitreous humor. It has been well established that drugs enter the vitreous humor as a result of passive diffusion from the blood (Felby, 1969), influenced by concentration gradient and lipid solubility.

The estimation of drug impregnation in vitreous humor, as well as in plasma and urine only reflects consummation during the preceding 1-3 days and the concentrations determined are strongly influenced by the dose and the time of the last exposition.

On the contrary, hair appears to be a particularly interesting substrate for the investigation of chronic drug abuse. Drug passes from circulating fluids into the hair and remains firmly bound there. Depending on hair length, it is therefore possible to trace the drug intake of an addict over periods longer than six months.

In this report, we present the comparison of post mortem drug of abuse concentrations in vitreous humor and whole blood, and the utility of hair samples for quantifying illicit substances.

MATERIALS AND METHODS

Vitreous humor samples (1-2 ml) were collected at autopsy (22 cases) with plastic syringues. Samples were stored at -20° C without preservative or hyaluronidase until being analyzed.

Hair samples (54 cases) of 30-40 strands, weighing at least 50 mg, were cut as close as possible to the scalp on the back of the head. The hair was decontamined by washing the specimens in 5 ml ethanol for 15 min at 37° C. The protein matrix of hair was destroyed by incubation in 1 ml 1M NaOH for 10 min at 100° C. Preparation was achieved by neutralization with HCl and centrifugation.

In both cases, drugs were extracted and quantified using standard GLC methods. Confirmation was obtained by mass spectrometry.

RESULTS AND DISCUSSION

1. VITREOUS HUMOR

22 cases suitable for study were available. Opiates, barbiturates and ethanol levels were determinated in blood and vitreous humor and the blood/vitreous humor ratios were calculated. Table 1 shows these results.

Table 1 : Drug blood concentrations compared with vitreous humor
(All concentrations are in mg/l)

drug	blood	vitreous	B/V	drug	blood	vitreous	B/V
secobarbital	14.64	2.06	7.11	morphine	0.02	0.35	0.06
secobarbital	1.27	0.10	12.70	morphine	0.17	0.03	5.77
secobarbital	42.17	14.80	2.85	morphine	0.46	0.47	0.99
secobarbital	3.24	2.30	1.41	morphine	0.96	1.38	0.69
amobarbital	2.79	0.24	11.60	ethanol	2320	2720	0.85
amobarbital	4.05	2.96	1.37	ethanol	190	240	0.79
vinbarbital	58.04	10.04	5.78	ethanol	1680	1720	0.97
pentobarbital	58.92	10.81	5.41	ethanol	1800	2160	0.89
codeine	1.02	2.60	0.39	ethanol	1020	1240	0.82
codeine	3.47	0.33	10.51	ethanol	840	980	0.82
morphine	0.56	3.70	0.15	ethanol	1520	1810	0.84

In all cases the drug found in blood was also found in vitreous humor. This study demonstrates that it is possible to make a toxicological screening in vitreous humor but this fluid is only interesting for qualitative informations. Excepted for ethanol, no correlation between blood and vitreous humor concentrations could be established. Therefore one cannot estimate lethal blood concentration with the only measure of vitreous level. On the opposite, it was found a significant correlation between blood and vitreous humor alcohol levels ($r = 0.99$) with a regression equation of : $B = 0.86V + 1.39$. The mean ratio of blood to vitreous humor ethanol concentration is 0.86 ($SD = 0.06$) which is consistent with theoretical considerations and previously reported data (Caplan 1990 ; Yip 1990).
In conclusion, vitreous humor appears to be a suitable fluid for a qualitative toxicological screening and can safely be used to estimate blood level in the only case of ethanol.

B. HAIR ANALYSIS.

The absence of drug metabolism in hair and the fairly uniform hair growth rate of about 1 cm per month may provide an historical account of drug use from analysis on hair samples. Hair is easily collected without trauma on the part of the donor, it can be stored without deterioration, and its contents can be analyzed with standard methods. Hair drug concentrations in 54 medical examiner's cases, including 22 post-mortem cases, are summarized in Table 2.

Table 2 : Concentrations of drug in human hair (ng/mg)

drug	level	drug	level
diazepam (1)	1.37	morphine (21)	0.29-27.10
nordiazepam (2)	1.04 - 1.47	codeine (3)	0.31 - 4.21
secobarbital (3)	21.6 - 58.9	benzoylecgonine (2)	1.21 - 3.41
phenobarbital (4)	21.7 - 137.3	amphetamine (3)	0.96 - 12.71
amitriptyline (6)	0.04 - 1.89	fenfluramine (1)	14.1
betaxolol (3)	0.6 - 2.8	cannabis (27)	0.40 - 2.70

( ) : nb of cases

Concentrations found in the hair were within the range of previously published evaluations (Püschel 1983 ; Ishiyama 1983 ; Marigo 1986 ; Baumgartner 1989). In the case of heroin abuse, codeine was also identified. The codeine could result from its presence in opium used for preparing heroin. In an other study, we have quantified nicotine in the hair of smokers and non-smokers (Kintz 1991). In the hair of 22 non smokers, nicotine content range from 0.06 to 1.82 ng/mg and there was significantly more nicotine in the hair of 42 smokers with a nicotine concentration range from 0.91 to 33.89 ng/mg.
It is of particular importance to specify the nature of the hair, since drug concentrations are different in the hair of the head, the axillea and the pubis. For example, the highest morphine concentration was found in pubic hair (2.46 ng/mg) followed by hair of the head (0.88 ng/mg) and axillary hair (0.40 ng/mg). These differences may be explained by the different growth and length of the hair.
Because head hair grows at approximately 1.0 $\pm$ 0.2 cm per month, the window of detection for hair analysis can range over many months to years. Furthermore, if the strands of hair are cut into sections (for example one month interval, i.e. 1 cm), one is able to obtain information on the pattern of an individual's drug use, that is, whether use is decreasing, constant or increasing (Fig. 1). This is also particularly interesting for monitoring compliance of medication intake.

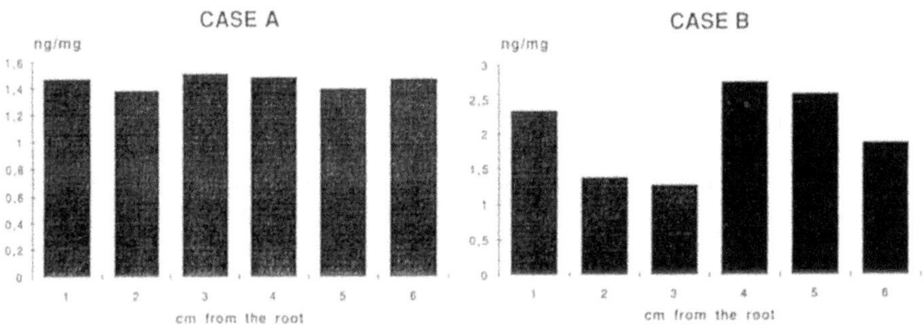

Figure 1 : Pattern of morphine use. A : constant use ; B : with variation.

Determination of gestional drug exposure by hair analysis is an other application of this technique. Neonatal hair from infants whose mothers were known to be heroin users ranged from 0.81 to 2.40 ng/mg of morphine. Hair analysis may identify intrauterine exposure to heroin in babies when a maternal drug history is not available or of doubtful thruthfulness.
During control tests of hair fragments, a drug addict is not able to hide the fact of drug abuse, even by deliberately abstaining for several days before sample collection, although this may produce negative response in urine analysis.
Hair analysis is, therefore, particularly useful for the screening of personnel in highly sensitive positions, like army, administration or aviation.
In summary, it appears that the value of hair analysis is steadily gaining recognition in clinical toxicology for evaluation of treatment compliance and in forensic toxicology for justice applications.

References :

Baumgartner WA, Hill VA, Blahd MD (1989) Hair analysis for drugs of abuse. J. Forensic Sci 34 : 1433-1453

Caplan YH, Levine B (1990) Vitreous humor in the evaluation of postmortem blood ethanol concentrations. J Anal Toxicol 14 : 305-307.

Coe JI (1969) Post mortem chemistries on human vitreous humor. Am J Clin Pathol 51 : 741-750.

Devgun MS, Dunbar JA (1986) Biochemical investigation of vitreous : applications in forensic medicine, especially in relation to alcohol. Forensic Sci Int 31 : 27-34.

Felby S, Olsen J(1969) Comparative studies of post mortem barbiturate and meprobamate in vitreous humor, blood and liver. J Forensic Sci 14 : 507-514.

Ishiyama I, Nagai T, Toshida S (1983) Detection of basic drugs (methamphetamine, antidepressants and nicotine from hair. J Forensic Sci 28 : 380-385.

Kintz P, Ludes B, Mangin P (1991) Evaluation of nicotine and cotinine in human hair. 43rd annual meeting of the American Association of Forensic Sciences, February 18-23, 1991 Anaheim, USA - J Forensic Sci, to be published.

Marigo M, Fagliaro F, Poiesi C, Lafisca S, Neri C (1986) Determination of morphine in the hair of heroin addicts by HPLC with fluorimetric detection. J Anal Toxicol 10 : 158-161.

Püschel K, Thomasch P, Arnold W (1983) Opiate levels in hair. Forensic Sci Int 21 : 181-186.

Vogel J, Hodnett CN (1981) Detection of drugs in vitreous humor with an enzyme immunoassay technique. J Anal Toxicol 5 : 307-309.

Yip D, Shum B (1990) A study on the correlation of blood and vitreous humor alcohol levels in the late absorption and elimination phases. Med Sci Law 30 : 29-33.

Ziminski KR, Weymyss CT, Bidanset JH, Manning IJ, Lukash L 51984) Comparative study in post mortem barbiturates, methadone, and morphine in vitreous humor, blood and tissue. J Forensic Sci 29 : 903-909.

Young Scientists Poster Award

The Young Scientists Poster Award is given to the first author of the best poster who must be younger than 35 years.

The winner is announced at the closing ceremony of the EUROTOX Congress. The abstract is published in the congress proceedings.

The 1991 winner was Monika Leutbecher, Dortmund.

ETHYLENE OXIDE: METABOLISM IN HUMAN BLOOD AND ITS IMPLICATION TO BIOLOGICAL MONITORING

Leutbecher, Monika, Langhof, Th., Peter, H., Föst, U.

Inst of Occupational Health, Ardeystr. 67, D-4600 Dortmund 1, FRG

Ethylene oxide is an important industrial chemical, used for the production of shampoos and detergents and for the sterilisation of heat-sensitive materials. In addition, ethylene oxide is an air pollutant emmitted by various sources like exhaust fumes and cigarette smoke and is metabolically produced from its precursor ethylene by mammalians. The carcinogenic potency of ethylene oxide has clearly been proven in animal studies. Therefore, a workplace exposure has to be monitored. Due to the importance of a determination of individual blood samples for biomonitoring, the metabolism of ethylene in blood and its influence on the rate of sister chromatide exchanges (SCE's) was determined.

Blood samples of volunteers were incubated with ethylene oxide. The decrease of ethylene oxide was determined in the gas phase. Blood samples of this group were also used to determine the SCE rate with and without in vitro exposure towards ethylene oxide. In additional experiments, freshly drawn blood samples of some of these samples (6 persons) were incubated with $^{14}C_2$-ethylene oxide. At intervals, 0.5 ml aliquots were isolated and the radioactivity in plasma, erythrocytes and lymphocytes was measured.

The results show two distinct groups of humans with significant differences in the metabolism in blood. While one group, so-called "conjugators", show a rapid decrease of ethylene oxide in the gas phase, a slow decrease of the substance was observed for the other group, so-called "non-conjugators". In the incubations with labelled compound, a highly significant difference in the conjugation of ethylene oxide in the erythrocytes was found, while there were only slight differences in lymphocytes and plasm. A separation of high and low molecular weight fraction of the erythrocyte cytoplasm (cut off: 10000 Da) showed a 8-fold higher conjugation of ethylene oxide to a slow molecular weight compound in conjugators than in non-conjugators.

This enzyme polymorphism observed for ethylene oxide in human blood may influence the cytogenetic toxicity of ethylene oxide in lymphocytes and the binding of the substance to blood macromolecules.

Subject Index

GPSR Compliance
The European Union's (EU) General Product Safety Regulation (GPSR) is a set of rules that requires consumer products to be safe and our obligations to ensure this.

If you have any concerns about our products, you can contact us on

ProductSafety@springernature.com

In case Publisher is established outside the EU, the EU authorized representative is:

Springer Nature Customer Service Center GmbH
Europaplatz 3
69115 Heidelberg, Germany

www.ingramcontent.com/pod-product-compliance
Lightning Source LLC
Chambersburg PA
CBHW061201240926
51846CB00006B/1087
9783642772627